W9-DFK-084

DAVID G. CAMPBELL

A LAND OF
GHOSTS

THE BRAIDED LIVES OF PEOPLE AND

THE FOREST IN FAR WESTERN AMAZONIA

RUTGERS UNIVERSITY PRESS

NEW BRUNSWICK, NEW JERSEY

Second paperback edition published by Rutgers University Press, 2013

Library of Congress Cataloging-in-Publication Data
Campbell, David G.
  A land of ghosts : the braided lives of people and the forest in far western Amazonia /
David G. Campbell.—1st American pbk.
    p.   cm.
  Originally published: Boston : Houghton Mifflin Co., c2005.
  Includes bibliographical references and index.
  ISBN 978-0-8135-4052-8 (pbk. : alk. paper)
  1. Indians of South America—Brazil—Acre—History. 2. Indians of South America—
Juruá River Valley (Peru and Brazil)—History. 3. Indians of South America—Juruá River
Valley (Peru and Brazil)—Social conditions. 4. Indigenous peoples—Ecology—Juruá River
Valley (Peru and Brazil) 5. Rain forest ecology—Juruá River Valley (Peru and Brazil) 6.
Ethnobotany—Juruá River Valley (Peru and Brazil) 7. Endangered ecosystems—Juruá
River Valley (Peru and Brazil) 8. Juruá River Valley (Peru and Brazil)—Social conditions. 9.
Juruá River Valley (Peru and Brazil)—Environmental conditions. I. Title.
  F2519.1.a27c36 2007
  981′ .1200498—dc22

2006035928
Copyright © 2005 by David G. Campbell
Reprinted by special arrangement with Houghton Mifflin Company

All rights reserved
No part of this book may be reproduced or utilized in any form or by any means, electronic
or mechanical, or by any information storage and retrieval system, without written permission
from the publisher. Please contact Rutgers University Press, 100 Joyce Kilmer Avenue,
Piscataway, NJ 08854-8099. The only exception to this prohibition is "fair use" as defined by
U.S. copyright law.

Visit our website: http://rutgerspress.rutgers.edu

Book design by Victoria Hartman
Maps by Jaques Chazaud

Manufactured in the United States of America

Praise for the hardcover edition:

"Only very rarely are [reviewers] sent something so obviously a work of real quality that all they can do is say: 'Buy this book!' David G. Campbell's *A Land of Ghosts* falls into this category. Campbell's masterful achievement is to distill a lifetime's experience, thought, and feeling into 288 pages in which every word is delivered with unerring grace. It is an instant classic."
—Toby Green, *The Independent*

"*A Land of Ghosts* is a sizzling pilgrimage—a rodeo of breath-catching details, kingdoms of fauna and flora and riverine skyscapes—haunted by trenchant foreboding that this topmost cream of creation will soon be destroyed."
—Edward Hoagland

"David Campbell went to the Amazon for science, and stayed for beauty. His book brilliantly evokes the beauty of the river basin, the plants and animals, and, most of all, the people. There's more too: the terrible underside of beauty, an environment disrupted and partly broken by human insensitivity and greed."—Chet Raymo

"To read this magisterial and heartbreaking book is to experience a sense of irremediable loss, not of symbols or of shadowy simulacra but of living, pulsing individual creatures in all their unimaginable variety."—*The New York Sun*

"A fluent and highly intelligent book."—Joe Kane, *Orion Magazine*

"Like the sinuous Amazon River he writes of so elegantly, ecologist David Campbell glides between travelogue and natural history."—*American Scientist*

"The characters are drawn magnificently—even the trees [David G. Campbell] describes have personalities. This is the book for extreme adventurers who are happiest at home in easy chairs reading about the challenges scientists face going where few have gone before."—*The Washington Times*

"*A Land of Ghosts* is science at its most moving, a small classic wrested from a vast but fragile wilderness."—*Men's Journal*

Books by David G. Campbell

The Ephemeral Islands

The Crystal Desert: Summers in Antarctica

Islands in Space and Time

A Land of Ghosts

A LAND OF GHOSTS

FOR ORIA

Contents

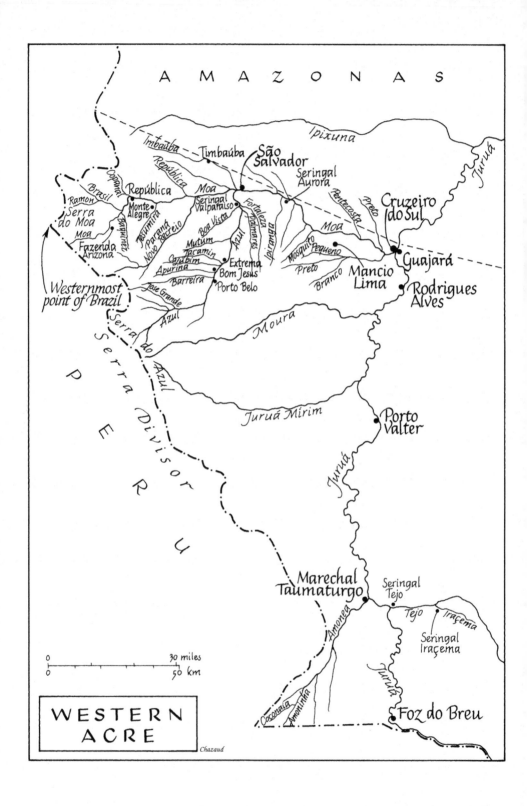

AMAZONAS

Ipixuna

Imbaúba

Timbaúba

São
Salvador

Seringal
Aurora

Cruzeiro
do Sul

República

República

Moa

Brasil

Copunai

Ramon

Monte
Alegre

Seringal
Valparaíso

Serra
do Moa

Moa

Bananeiras

Tejumirim

Paraná
Novo Recreio

Boa Vista

Fortaleza

Palmares

Ipiranga

Azul

Moa

Pentecosta

Preto

Guajará

Fazenda
Arizona

Mutum
Tacamin

Cajubim

Apurinã

Barreira

José Grande

Extrema
Bom Jesus

Porto Belo

Mosquito
Pequeno

Preto

Branco

Mancio
Lima

Rodrigues
Alves

Westernmost
point of Brazil

Azul

Serra do Azul

Serra
do Azul

Moura

PERU

Serra Divisor

Juruá Mirim

Porto
Valter

Juruá

Marechal
Taumaturgo

Seringal
Tejo

Tejo

Iraçema

Amônea

Seringal
Iraçema

Juruá

0   30 miles
0   50 km

Coconaia

Amôninha

Juruá

Foz do Breu

WESTERN
ACRE

Chazaud

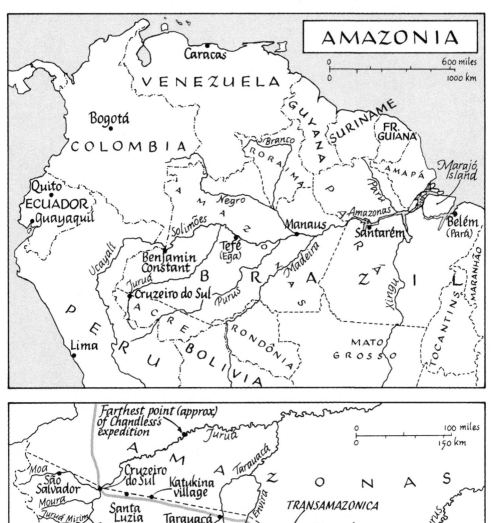

AMAZONIA

0 — 600 miles
0 — 1000 km

Caracas

VENEZUELA

Bogotá

COLOMBIA

GUYANA
SURINAME
FR. GUIANA

Branco
RORAIMA

AMAPÁ

Quito
ECUADOR
Guayaquil

Negro
Solimões

Marajó Island

Manaus
Amazonas
Santarém
Belém
(Pará)

Ucayali

Benjamin Constant

Tefé
(Ega)

B

Cruzeiro do Sul
ACRE

Juruá

Purus
Madeira
Xingu

A
Z
I
L

MARANHÃO

P
E
R
U

Lima

BOLIVIA

RONDÔNIA

TOCANTINS

MATO GROSSO

ACRE

0 — 100 miles
0 — 150 km

Farthest point (approx)
of Chandless's
expedition

Juruá

A M A Z O N A S

Moa
São Salvador
Moura
Juruá Mirim

Cruzeiro do Sul

Katukina village

Tarauacá

Santa Luzia
Tarauacá

TRANSAMAZONICA

Envira

Purus

Porto Valter
Marechal Taumaturgo

Juruá

Tarauacá

Muru
Paraná do Ouro

Manuel Urbano

Purus

Caeté

P
E
R
U

Foz do Breu

Jordão

Envira

Purus

C
R
E

Porto Acre

Iaco

Acre

Rio Branco

Xapuri

Chazaud

BOLIVIA

A LAND OF GHOSTS

PROLOGUE
The Odyssey

The reality of yesterday becomes a fable
. . . and one forgets it.

—Paul Gauguin

THIS IS A STORY from the edge of the human presence on our planet, the far western Amazon River Valley, where the forest envelops the horizon and the sky vaults in indifference to our small ways. It is a tale of men and women on a frontier so vast that they seem eclipsed by it. It is the story of their survival and their despair—and sometimes their triumph. It is a record of scientists who seek understanding of the natural world and of Native Americans who are losing that world, as their age-old culture—and their cosmos—disintegrates.

For thirty years I have conducted ecological studies in the Brazilian Amazon, in particular around the remote headwaters of the Rio Juruá, in the state of Acre, near the Peruvian border. In the course of my journeys there, I have become friends with—indeed, part of the extended family of—the pioneers along the rivers and highways.

A wanderer from another continent, I have shared their lives and for a moment was captivated.

I lived among four cultures: the recently arrived colonists from Brazil's densely populated east who have settled along the Transamazon Highway; the Caboclos, people of mixed heritage who are masters at making a living along the rivers in this intractable land; the local Native Americans; and the university-trained scholars of the Western empirical scientific tradition.

The colonists settle along the moth-eaten edge of the highway—the Transamazonica—growing manioc and coffee and raising a few pigs and chickens in the sandy soils. Some bring the inappropriate farming technologies of Brazil's arid east to the tropical rainforest; or, worse, some are city folk with no understanding of farming or the soil. The colonists inevitably fail; the forest succumbs to ephemeral small fires; its fertility is transient.

The Caboclos are fiercely defensive of their lonely environment. Of African, European, even Middle Eastern descent—the vestiges of past diasporas—they are mostly Native American. They all speak Portuguese now. Although most have lost their native languages, they have not necessarily lost all their native skills. They know how to eke out a living in the river and forest. Still, they are not masters of their environment. They are seduced by the forces of commerce, especially by rubber-tapping and cattle-ranching.

The Native Americans, who once understood every nuance of this forest and had a name for every one of its species, have become dislocated and confused, coveting Western ways but unable to grasp them. In a generation many will become Caboclos themselves, but in the quantum jump from Native American to Caboclo they will lose their traditions, their culture, and their very language.

Each family along the rivers has its own tale, often tragic, sometimes heroic. The best interests of the different groups are often in conflict in a land where there is no law, a land neglected by an indifferent government hundreds of miles away. As in the North American West of more than a century and a half ago, conflicts in western Acre today are settled by gunfights and knives; feuds endure for generations.

I came to the River for science—to explore one of the great and enduring conundrums of nature: how can so many species coexist in such a small patch of this earthly orb? My colleagues and I try to decipher how ecosystems evolve and are maintained. I suppose, in retrospect, that my particular discipline, botany, was irrelevant. I could just as easily have been an entomologist or an ichthyologist. In any case, this lonely terrain would have seduced me into returning again and again. For me, what was important was to bear witness to this place that is like no other, to be a player in the game of science, and to have a dalliance with knowing—aware all the while that the paradigms of today will be supplanted by better ideas tomorrow.

How privileged I was to experience this place at this moment in history. Had I made this journey 150 years ago, the vistas and ways of life would have been the same, although my tale would have had more danger. As I do today, I would have reveled in the dumbfounding diversity of life forms. Unlike today, however, then I could have done little more than pose questions.

One might imagine that it's a simple goal to understand the way things are assembled on this planet, the rules of order. Yet the measurement of patterns of species abundance and rarity—and the slow realization that this diversity was generated not by a benign and constant environment but by disturbance and inconstancy—has been one of the great intellectual quests of our time. How many species

A LAND OF GHOSTS

can evolve in a small space and be held there? What are those species? Can we ever count them all, describe them? Are they all variations on a limited number of themes, or are they a richly diverse mélange? Why are there any rare species at all? Why haven't the most successful species pushed aside the others?

The field work necessary to answer these questions is laborious: every tree and shrub on each small plot must be collected, identified, and mapped. The task is all the more difficult because many of the species in the area I study are new to science, or have been described only in an arcane monograph in a specialized library. I now have data from a total of 18 hectares. A hectare is an area of only 100 by 100 meters, and 18 hectares is just a mote in the vast Amazon Valley. But that small area seems a universe to me. Its diversity is stunning: more than 20,000 individual trees belonging to about 2,000 species—three times as many tree species as there are in all of North America. And each tree is an ecosystem unto itself, bearing fungi, lichens, mosses, ferns, aerophytes, orchids, lianas, reptiles, mammals, birds, spiders, scorpions, centipedes, millipedes, mites, and uncountable legions of insects. The number of insect species alone—especially of beetles—may exceed the combined total of all the other species of plants and animals in the forest.

From repeated observation and from the patient teaching of the Native Americans and the Caboclos, I have learned to regard each tree on my study sites as an individual: I know its indigenous name, the lore that surrounds it, the medicines that can be made from it, its uses for food or fiber, and myriad other arcane items of its natural history. I know its hitchhikers, the traplining bees that are its pollinators, the bats that disperse its seeds, its parasites, and the lianas that steal its light. I know how the sap smells and feels; how the ocelots have raked its bark and scent-marked it with their urine.

I know the generations of pacas that live under its flying buttresses, the army ants that bivouac every few years in its hollow trunk, the tinamou that forlornly flutes from its highest bough. I've observed its seasons of poverty, when the land is unyielding and sparse, and of productivity, when the red-footed tortoises and white-lipped peccaries eat its fruits. The story of each tree is a story of seasons of birth and death, a story as old as Earth herself.

To walk through the forest for a day or even a month is like taking a snapshot of a child and expecting to know what she will look like as an adult or how she appeared as a baby. And so I returned each year to my little patches of forest to learn the destinies of the trees I had so meticulously measured five, ten, or twenty years before. Had they survived? How much had they grown? Were the diversity and species composition of the forest changing? These simple questions, so basic to our understanding of pattern and process, had no obvious answers. I learned that a tropical forest cannot be measured in only three dimensions. Time—the inscrutable fourth dimension that we measure with clanking mechanical wheels or the vibrations of molecules—is the key to understanding how a tropical forest was wrought, how its diversity is maintained, and what it will become. When I first visited the Amazon, the trees seemed to be immortal; by comparison my life seemed as transient as an ant's. But I learned that the vast majority of the trees in my forest plots died as seeds or sproutlings, and a sapling was lucky to live to reproductive age. Some trees were broken by tipsy neighbors or suffocated in the lethal shade of the light-greedy canopy; then their minerals were recycled into new recruits. Today these trees exist only as numbers in my data book. Only a few saplings had the good fortune to sprout in a light gap created by the death of a neighbor, allowing them to invest in a trunk and the woody infrastructure necessary to lurch into the canopy, stake a claim, and win a place in the sun. These few will become the giants that spread their boughs and cast their seeds over all the others, and I'm sure they will stand long after I have vanished.

The more I explore the rainforest, the more questions I have and the more ignorant I feel. This, of course, is the signature of good science. True, I have identified and mapped every tree and shrub in some small nooks and corners of this vast forest. I understand these microcosms. But my research is constrained by its small scale, which doesn't reveal the whole any more than a rain puddle is representative of a summer pond full of living things.

In 1974 I was a member of the first team of botanists to collect plants along the brand-new road being constructed between Manaus and Porto Velho, part of the Transamazon Highway system. The road, intended as a sinew of commerce, was the portal for my entry into the forest, into places that had been inaccessible until then. The highway ran straight through the forest, indifferent to rivers and terrain, across highland and floodplain. At first Brazil's military government kept the settlers and squatters at bay, and for a while the trees grew right to the edge of the road. At the end of the day we would simply pull the Jeep over to the edge of the road and make camp. Our campsites were full of wildlife, because the new road crossed the territories and migratory pathways of the forest animals. One night, near the Igarapé Mutum, we had to sleep in the Jeep, because a jaguar prowled our camp, coughing and scent-marking the Jeep and our boxes of supplies with urine. For weeks the Jeep and our clothes smelled like a cat's litter pan.

The purpose of our expedition was to record the plant species in the area before the road brought in farms and cattle ranches. We collected plants from dawn until dusk and spent the evenings by the campfire pressing and drying them. In the parlance of botar.. cal collecting, this unglamorous work is known as hay-baling. Yet it was hardly routine: about one percent of our collections were species new to science. Since we collected between fifty and a hundred plants per day, we had as many as seven new species each week. By contrast, the last new tree species in North America—and probably in Europe, too—were collected and identified more than a century ago. Our discoveries were as original as any during the heyday of Victorian biological exploration a century before.

Earth is four and a half billion years old. She will survive another five billion years before the corona of her star expands and envelops the inner, rocky planets. How is Earth doing in her middle age? Pretty well, I'd say. On other planets of the same age life has not yet evolved, let alone intelligent life. They are as sterile and inarticulate as cinders.

Some places on my natal planet are more voluptuous than others. The Amazon River and its forest are the last continent-sized stretch of untrammeled wilderness on Earth and the greatest expression of her biological diversity. The Amazon Valley, in fact, has more species than have ever existed anywhere at any time during the four-billion-year history of life on Earth. Yet just at this moment of peak biotic eloquence, we know that inevitably most of it will be lost. Over the next several decades Earth will lose hundreds of thousands—perhaps millions—of species. Most of this extinction will occur in the tropical forests and especially in Amazonia. In the course of a generation or two, the Amazon rainforest will be destroyed both as a wilderness and as a functioning ecosystem. It is an old story, stale news by now. Yet the loss of this place, I'm convinced, will be the most egregious event of a generation of atrocities.

Much of the deforestation of Amazonia is being carried out by hopeful innocents. They—the heroes of this book—are pilgrims on a frontier as wild as any that humans have known. This is not a tale of villainy, like the chronicles written by the ambulance chasers after the assassination of Chico Mendes. Instead, it has all the trappings of a classic tragedy: good intentions gone awry, heroes who don't know any other way.

The forest along the highway from Manaus to Porto Velho is gone now, having retreated over the horizon on either side of the road. The Igarapé Mutum has become a wasteland. After the road crews departed, the squatters invaded, burning the forest and setting down hopeful little plots. Cattle ranchers, given massive tax incentives by the government, came too. Some of the plant species we collected have never been seen since, and it is unlikely that they ever will be seen again. Our collections will be the only evidence that they existed.

I am confident that my species has the vision and discipline to prevent this wholesale destruction. Perhaps this book will offer a few morsels of hope by showing that there are people, living far from the highway, who know how to survive in the forest without destroying it. But the forces that want to consume the forest are overwhelming. During the next hundred years Earth's human population may be too voracious to allow much wilderness to survive. Even if we temper our appetites, only a few scraps of wilderness, each as isolated as Central Park, will remain. We will have forever denied our descendants the chance of living an adventure like the one I describe in this book.

I am no futurist, but I accept that the loss of the Amazonian forest will deplete the soils, create worldwide changes in climate, and result in an extinction of species as great as that at the end of the Cretaceous Era, sixty-five million years ago, when the indifferent heavens—a collision with an asteroid or comet and the subsequent darkening of the sky with smoke and dust—caused about 50 percent of all species to disappear. Life on Earth, I'm sure, will eventually survive the human catastrophe, too. Earth is a forgiving mother with a long memory. Yet after the Cretaceous collision it took ten or fifteen million years for new players to evolve and replace those that were lost. And even then the new life forms consisted of variations on a diminished number of themes; today there are no ammonites, no dinosaurs. Human history is but a microsecond on Earth's time scale; as far as we're concerned, we are changing the world forever.

Perhaps every planet with intelligent life must endure this tragic and irreversible adolescence, when her children run amok. After all, as we lose our biological heritage we acquire new traditions. We are now developing a godlike knowledge of physics, from subatomic particles to quasars. Artificial intelligent life is just over the horizon. We know how to enslave the DNA from just about any organism to work for us in the simple and innocuous cells of bacteria. We are making new life at the same time we are destroying the old forms.

My profession has made me a recorder of events that will change humanity and, for a while, the world. Should my descendants in future centuries read these words, they may find them a strange tale indeed, of a time and a place whose life forms are mostly alien to them.

These days I am drawn to Homer, who wrote of voyagers along the shores of the eastern Mediterranean when the land was mantled in forest and prowled by lions, leopards, and wolves, when the sea was graced by the songs of sirens. Describing an island in the Adriatic, Homer wrote:

> There is a wooded island that spreads, away from the harbor, . . . for-
> ested; wild goats beyond number breed there, for there is no com-
> ing and going of human kind to disturb them, nor are they visited by
> hunters, who in the forest suffer hardships as they haunt the peaks of
> the mountains, neither again is it held by herded flocks, nor farmers,
> but all its days, never plowed up and never planted, it goes without
> people and supports the bleating wild goats.

Those vistas, beasts, and sounds seem improbable today, when the
Turks and Greeks eke out a living on a shaven and bruised terrain
where only the spiny and sharp-scented herbs disdained by sheep
and goats have survived, and where the teratogenic sea yields lumpy,
toxic fish and three-limbed pelicans. The people of the Mediterranean have forgotten the ancient tapestry of their land. Their sea
runs empty and silent.

Today we know that Homer's singing sirens were humpback
whales, now extinct in the eastern Mediterranean. And we know
that lions once ranged as far north as central Europe (indeed, Europe had its own subspecies, distinct from those of Africa and Asia).
By Homer's time the lions had already been hunted to extinction;
as enemies of the "herded flocks" of goats and sheep, they were the
first to succumb to the shepherd's bow and lance. Yet Homer men-
tions lions sixty-two times in his poems; lacking the real creatures,
the blind old poet subsumed them into myth, symbols of strength
and courage.

Will the Amazon be like the eastern Mediterranean someday?
Will the people lose their way and forget the species they once
knew? If read a century hence, will my odyssey seem as unlikely as
Homer's? Am I, like the blind old poet, writing a future myth?

For sure, Amazonia has its improbable beasts, its own myths.
Homer would have delighted in the stories of six-meter-long ana-

condas, of *angelim* trees as tall as the Colossus of Rhodes, of mythical *mapinguarí*, a beast with backward-facing feet (it may prove to be a not-yet-described ground sloth). There are plenty of dark and improbable sights in Amazonia—monsters to outsiders, yet commonplace neighbors to those who know the area. But as in Homer's Greece, there is a dark inevitability here. All of this beauty—this refuge of the imagination—will end soon. In a few years this forest will succumb to flames, the mediocrity of mono-cultures, nationalistic paranoia, and the grubby quest for gold (a tiresome obsession as old as the New World). My generation will be the last to live in a species-rich world, in a time when most taxa remained to be discovered. And my generation will watch that world end. Our own species is forging the next great earthly ex-tinction, diminishing forever our only homeland. Perhaps never before in Earth's history (barring extraterrestrial impacts) have the events of a few decades been so important. The changes being wrought in Amazonia will alter the trajectory of life on Earth. The decisions that we make now, at the cusp of two millennia, will have reverberations five hundred years from now, and five million years.

At least the Greek words for lions and sirens lived on after the beasts went extinct. Indeed, the Greek language became the tem-plate of Western tongues. The native Amazonians, though, are los-ing their cultures and languages at the same time they are losing the biological diversity of their world. The Native Americans have for-gotten thousands of the words they once used to describe this for-est. It has become a forest without cognizance. Species and the words for those species are going extinct. Both life and extinction have be-come anonymous. We scientists are forced to decipher pattern and process in a place where most species remain undescribed and their functions unknown. Now we must rename its parts, construct a new scaffolding of taxonomy. We are mapping a New World—using expired Greek and Latin words from another time and another continent as our tools.

🐚

I came to the River for science, but I stayed for the beauty. My memories of the species I found—each an invocation of sunlight and water and minerals—and of the play of light in the canopy, the night sounds, the aromas and textures of the forest, the time and space shared with friends on the frontier make up a tapestry of experience so rich that now, years later and thousands of kilometers away, it imbues my papery life with dimension and perspective.

I once wrote a book about Antarctica, a place where form and light are distilled into a few simple, evocative phrases, where only a few species have managed to climb ashore and survive. Antarctica is parsimonious and therefore easy to characterize. It is biological haiku. But how do I describe the inchoate green tapestry of the Amazon Valley, this apex of earthly diversity? Imagine: there are more species of lichens, liverworts, mosses, and algae growing on the upper surface of a single leaf of an Amazonian palm than there are on the entire continent of Antarctica. How do I reduce this voluptuous diversity to words?

Where do I begin?

1

SOUTHERN CROSS

Enter these enchanted woods,
You who dare.

—George Meredith

MY JOURNEY BEGINS in the frontier outpost of Cruzeiro do Sul,
one of Earth's forgotten places. The only town of any significance
in western Acre, Cruzeiro lies on the right bank of the Rio Juruá
atop seven low hills, the first reverberations of the Andes in the pla-
nar sediments of lowland Amazonia. The town's name, Southern
Cross, refers to the constellation that guided the fifteenth-century
Portuguese explorers below the equator and that on clear summer
nights etches the southern sky, its base always pointing toward the
South Pole. But a starry horizon is a rarity here; the forest intrudes
everywhere, and for a third of the year it rains. To see the night sky,
you have to step away from the few meager street lights to the margin of the river or into the clearings made by cattle ranchers on the
edge of town. But when you do see the sky, it looks as if heaven
were crashing to earth in a thousand shards of starlight.

The Rio Juruá at Cruzeiro is only 72 meters above sea level, and not a single rapid—let alone a waterfall—interrupts its passage to the Amazon River, which descends to the Atlantic Ocean, a journey of 4,500 kilometers, at the almost imperceptible rate of 1.5 micrometers per kilometer. This is the plain of lowland Amazonia, the greatest terrestrial wilderness on Earth, with a thousand tributaries wandering in aimless meanders and vast swaths of forest that have not been explored by scientists.

For both the vanished Incas and the modern-day Brazilians, the western Amazonian forests were the end of the world, the place where civilization dissipated. Today it is hard to imagine why anyone bothered to establish a town here. There is no particular virtue to this spot. The river margin is neither stable nor especially fertile. The journey here is often dangerous and always tedious.

Cruzeiro do Sul's reason for being is rubber; the town is surrounded by the best wild rubber trees in the world. A century ago an ambitious young man with a bit of luck could get rich in Acre just by tapping rubber for a few years or, better yet, by becoming a *regatão,* an itinerant trader who bartered salt, gunpowder, coffee, and other necessities for *bolas,* the big balls of smoked rubber that were the most important article of commerce in Amazonia. These pioneers referred to the creamy latex of the Amazon rubber tree as *ouro branco*—white gold. The rubber tree is endemic to the Amazon, and for a while in the nineteenth century Brazil held the monopoly on a product that the industrializing world wanted desperately for making gaskets to seal steam engines, tires on which the newly mobile civilizations to the north rolled, and the newfangled prophylactics that allowed sex to be both safe and exquisitely sensuous. Now Amazon rubber trees have been introduced throughout the tropics, from Liberia to Borneo, and synthetics are being manufactured for just about every product in which natural rubber was used. The economy of elastic things has moved on, and Cruzeiro has become an economic atavism, an outlier of history.

Opposite Cruzeiro do Sul, on the left bank of the river, is the terminus of the Transamazon Highway, which begins 3,000 kilometers to the east in the gold-mining town of Maraba. For a few years after the Transamazonica was completed in 1973, Cruzeiro flirted with the industrial coast of Brazil. In those days as many as thirty trucks a year came to the town, laden with the delights of the great world beyond. As a trading center the town prospered, and for a while it became a boom town. But today the highway is washed out, its bridges rotted and collapsed, and Cruzeiro is once again isolated in the wilderness.

The Brazilian airline Varig has several flights a week to Cruzeiro from Rio Branco, the capital of Acre, 700 kilometers to the southeast. The planes carry just enough fuel to get here and back, which means that once committed to the return flight, they must fly through any weather, defying the dark thunderheads of the rainy season. The planes are always full, and sometimes you have to wait several weeks to get a seat. Residents return with all manner of glitzy items from the big city stuffed into suitcases: I have watched them carry on board scooters, bassinets, sewing machines, even portable generators.

Now that the Transamazon Highway is closed, most bulky cargo arrives in Cruzeiro do Sul on barges, which plod day and night southwest from Manaus for twenty-eight to thirty-two days, depending on the water level. The journey seems to take a lifetime. The traveler begins on the Rio Negro, whose waters are the color of strong tea, drifts south 15 kilometers into the caramel and chocolate swirls of the Rio Solimões, then pushes against the current for 500 kilometers to the mouth of the Juruá, where the river is confounded by a maze of anastomose canals, sandbars, and islands that change according to the whims of season and rainfall. In the dry season these sandbars heap logjams of trees ten meters high and a kilometer long; you sense violent and disturbing events ahead.

Then for three more weeks, the traveler must negotiate the countless braided meanders of the Juruá, where the sun is at one moment in your face and a few minutes later behind you. Sediments from the Andes are pushed in moving ridges over the river bottom, like washboards on a road, and the surface is fretted with ripples that belie the terrain below. In some places unseen obstacles create upwellings that appear like ripe boils, smooth-skinned and swollen, when they reach the surface. Over the millennia, the river has coiled through this forest like a slow-moving anaconda, changing its course by a hundred kilometers in as many years, leaving cays and isolated oxbow lakes in its wake. The lakes are islands of water in the forest; the cays are islands of forest in the middle of the river. These islands—real and virtual—are places of isolation and therefore of genesis, where the astonishing biological diversity of the western Amazon has evolved.

There is no clear and distant perspective on the Juruá, just the close forest and the occasional small settlement of rubber tappers: a few huts thatched with the leaves of *buriti* palms, perhaps the startling purple canopy of a flowering *jambu* tree. On the crumbling, waning side of the meanders is the *terra firme,* the dry upland forest, with Brobdignagian trees sixty meters tall. Their shallow roots are undercut by the restive river, and eventually the trees topple into the water, carrying with them whole communities of vines, strangler figs, orchids, ants' nests, wasps' nests, termitaria, bromeliads, ferns and philodendrons as tall as a man, along with the leaf litter and soils that have accumulated in their aerial roots. Sometimes when a tree topples, it seems as if a whole chunk of land has fallen from the sky. On the opposite bank, where the sediments accumulate, are the tiered forests of regeneration: the first wispy willows, then the broad-leaved cecropias, and finally a wilderness of viny *várzea*—the seasonally flooded forest—which is almost as grand and as tall as the terra firme. In the várzea the water rises fifteen meters every rainy season, sweeping away the understory and coating the tree trunks with gray mud. Nobody would dare to build a house there.

As I write these words, I am sitting at a spindly metal table at the Restaurante Flor de Junho, an outdoor tavern in the heart of Cruzeiro on the Praça do Triunfo, the Plaza of Triumph. I am quaffing a tall Antarctica beer—*estupidamente gelada*—with Arito Rosas, a mammalian paleontologist from the Federal University of Acre in Rio Branco. Arito is a natural historian, trained in both Western and indigenous systems of classification, who knows the fauna of Acre better than anybody else. He is carefully considering a paper wasp sipping at a puddle of beer on the plastic tablecloth. The jukebox is playing, in English, "Stop in the Name of Love." The bar has a few bottles of liquor strategically placed to catch the eye. They are for status only, since nobody can afford to buy them. This is a beer joint. Behind the bar is a portrait of a jaguar painted on velvet, a fan with smashed flies on the leading edge of its blades, and a fire extinguisher—the only sign of government authority.

We are watching the sunset as the day's last black vultures drift over the darkling riverfront and the evening's early bats, groggy with hunger after the long day's fast, careen after the summer moths. A cat, as black as a panther, is scrutinizing a rat among blue-flowered railroad vines in the vacant lot next door. The rat is not afraid; it is bigger than the cat.

Cruzeiro do Sul is a raucous city. Drivers here believe that an un-
muffled vehicle is more economical than a silent one, and an un-
bridled engine is considered a declaration of wealth and achievement. The town is coming to life after the stifling heat of the long
afternoon, when people retreated to their hammocks and the taxi
drivers pulled their cars into the shade to take a nap. The big four-
ton trucks that haul cattle and rubber to the waterfront are parked
around the square. Their drivers will spend the night enjoying the
street life, then sleep in hammocks strung between the fenders. Their
cabs are painted with sorrowful saints and comforting Madonnas;
plastic Disney characters hang from the rearview mirrors. Most of
the trucks were driven here over a decade ago, when the Transamazonica was open. Now they have little purpose; western Acre has
only 150 kilometers of paved roads.

The bar is shaded by a spreading *samaúma* (silk-cotton) tree,
a jambu tree, a *Terminalia* tree with platelike leaves, and a pink-
flowered *Calliandra,* from which a tyrant kingbird, perched on the
highest twig, surveys its squalid little territory. It leaps into the air
above the square—dipping, then rising—grabs a moth, and settles
again on the *Calliandra* to thrash it to death. A little dust cloud and
a broken moth wing drift down to the street.

The bases of the tree trunks are whitewashed. One has etched
into it: EURICO AMO IVANILOY, a declaration of adolescent passion
to an indifferent world.

On the street in front of me is a roving pack of *peixotes*—street
urchins—dressed in grubby T-shirts, shorts, and sandals, carrying
painted wooden shoulder packs full of gum and shoeshine imple-
ments. Probably one of them is the lovesick Eurico. Their T-shirts
bear little shards of thought, written in glitter paint in the language
of a distant land: *Ocean Pacific Wave Crazy; Hands That Make
Wishes; Hearts in Love, Brother of the Sea; Wake Up Your Ocean;*

Spirit of Louis. The peixotes forage for customers among the sidewalk bars and stands that sell *churrasco* kebabs, the white lightning known as *cachaça,* and cigarettes. Three young swaggerers gather around one of these wooden stalls, strutting with the exaggerated movements of first testosterone. They are trying—with little apparent success—to impress a teenage *gatinha* in a bright turquoise Spandex tank top, buying cigarettes at the Madrugá Kiosk. Is she femme Ivaniloy? She slouches ever so slightly, cigarette drooping from her lips, emulating the haggard insouciance of a middle-aged woman.

The *praça* has all the conveniences of city life: restaurant stalls, mechanics, canoe builders, the Sapateria Beira Rio (Riverside Shoe Store) and the Prótese Tiradentes (Prosthetic Teeth). On the east side is a white church with the signature red cross of Hermão José, an itinerant preacher, black-bearded and wild-eyed, who drifted through Acre almost forty years ago, inducting people into the throes of a wild and passionate form of Christianity. A block away, beyond the praça, is the eight-sided Cathedral of Nossa Senhora de Glória, the city's patron saint. Built of baked red bricks made from the Amazonian mud, the cathedral is always cool inside and strangely breezy. The Flor de Mayo (Mayflower) and Ganzo Azul (Blue Goose), the city's two principal bordellos, are just down the street, so clients can conveniently cleanse their souls between bouts of sin.

Downhill is the shop of the coffin maker. The sign reads:

EMPRESA FUNERÁRIA
SÃO JOÃO BATISTA
ATENDIMENTO DIA E NOITE

Funeral Parlor of Saint John the Baptist, Open Day and Night...
a good business, I would think. The atrium is stacked with vertical
coffins, dilated where the arms of the deceased would be crossed.
Most are child-sized. Made of raw lumber, they are not built for
permanence.

Farther toward the riverbank is the *bairro flutuante,* a floating
shantytown perched on spindly stilts or on huge pontoons of *assacú*
logs over reechy mud flats. The bairro has no sidewalks, only elevated planks, and it is self-cleaning: the garbage and sewage that
accumulate during the dry season wash away in the first flood of
autumn.

Tarzan lives with his wife, Geni, in the bairro. His true name is
Francisco Barbosa Derzi, but the locals had a hard time pronounc-
ing "Derzi" and transmuted it to Tarzan, whom they had read
about in comic books. His house, painted orange with a tin roof, is
on a walk appropriately named Lagoa. The heart of the house, indeed of the neighborhood, is Geni's kitchen. From dawn until sunset her red porcelain butane stove broadcasts the seductive aroma
of beans and rice and, on a good day, a simmering hen from the
coop out back. Tarzan owns a ramshackle wooden kiosk where he
sells batteries, tinned corned beef, cigarettes, and cachaça. In the
evenings, when the lambent light of the kerosene lanterns beckon
camaraderie, his shop becomes a place of drink, gossip, and the
good aroma of cigarettes hand-rolled from *coringa* tobacco. At
these moments Tarzan's shop seems to be an island of light in the
dark and unknowable presence that laps at the edge of the bairro.
Even though he spends most of his time in town these days, Tarzan
always knows what's going on in the interior. His customers—rubber tappers, smugglers, *pistoleiros,* and wanderers newly returned
from the forest—weave wonderful tales about their adventures.

They know where the malaria is, where the safe places are. This savvy—known in Brazilian Portuguese as *jeito*—can save your life.

Now and then the forest sends little hints of what is out there: the musk of a tapir that careened through the edge of town one day, the king vulture that spent an afternoon perched on the rooftop of the fish market. Last month hundreds of thousands of Uraniid moths—swallowtailed shards of black tinsel—blew into town. Their migration took two days to pass through, and a hundred died under every bright window, attracted to the unanticipated light. Before the moths there was an invasion of millions of yellow scarab beetles the size of corn kernels. You couldn't walk across the street without crunching them under your feet. Why? Had something in the forest gone bonkers on those days, causing it to expel its children? I suppose Cruzeiro do Sul and the surrounding roads and cattle ranches have disrupted the animals' migratory pathways, old corridors that survive in the inappropriate memory of instinct, which pass straight through the Praça do Triunfo.

Sometimes the forest releases multitudes of beasts. The men who gather in Tarzan's kiosk still talk about the evening of the full moon in July 1972, when Cruzeiro was invaded by a band of about four hundred *queixadas*—white-lipped peccaries. It was the annual festival of Corpus Christi, during which a procession of men, women, and children dressed in white bear an effigy of Christ through the Praça do Triunfo to the cathedral. The peccaries first appeared in the early evening along the forest margin near the hospital, then surged down the swale behind the brick factory, clacking their jaws in characteristic fashion. They swarmed the house of Arito's cousin Ivanilde, leaping at the door and breaking through a shuttered window. Screaming, she retreated into the kitchen and climbed on top of a table. "Ivanilde is a heavy lady," Arito explained, "and she nearly broke the table." Before fainting, however, Ivanilde killed a peccary by dropping an iron on its head. Moments later, her husband Antonio returned home with a shotgun and killed three others with a single blast.

By then Arito had called the men of the town to arms.

"I killed nineteen," he said, "four with a single shotgun blast. My cousin Jaí killed thirteen; another cousin, Ailson, killed twenty-six."

Yet the peccaries continued, spilling into the church and the school. Eventually, they tried to cross the Juruá, but they became confused, piling on top of each other and foundering.

"We pursued them even there, in our canoes, striking them on their heads with our paddles and drowning them."

In all, more than three hundred peccaries were slaughtered in Cruzeiro do Sul that night. Arito and his friends, their faces and shirts stained red, went straight from the massacre to the torchlight Corpus Christi procession, carrying the dead peccaries on their shoulders. For a week all of Cruzeiro feasted. Salted slabs of peccary meat were laid out to dry in the sun on planks by the beach. And within a few days the town was filled with black clouds of flies.

In most cities you can learn more about local life by visiting the market for a few hours than you can in the next month. But the municipal market of Cruzeiro do Sul is an inarticulate place that teaches you almost nothing about the region, selling instead the imported items that Acre can't produce: peppercorns, garlic, bags of pearly white rice, countless sacks of beans. Nearby the baker uses wheat flour imported 2,000 kilometers from southern Brazil. In this land, where manioc flour is the indigenous staple, the aroma of baking bread is the smell of conquest.

To understand Cruzeiro do Sul, you must walk down the sloping road to the broad, tawny beach. On this day in late August, one of the last days of *verão* (summer), the beach has become an emporium. A family of Kampas Indians, all wearing tightly woven cotton serapes, are camped around a fire of driftwood. A pair of lovers are seated on a log, nuzzling each other. Little boys playing *futebol* in the soft sand pause to watch dreamily as a canoe disappears around the first bend of the river. The beach is heaped with the offerings of the river and forest, brought to market in a hundred canoes: curly dried *pirarucú* fish, oil-sweating bolas of raw rubber (which smell of the *mata mata* wood fires over which they were smoked), coagulated slabs of *caucho* rubber, pastel mounds of *pupunha* palm fruits, and stacks of rough-hewn timber that perfume the waterfront with the sweet aroma of fresh sap.

Most of the canoes are dugouts, gnawed by fire from single logs. Their hulls, painted every imaginable color, have low transoms that nearly touch the water, but the prows are high and competent. Some of the dugouts are broad-bellied houseboats, complete with chicken coops and vegetable gardens planted in old broken canoes atop their ragged thatched roofs. The biggest ones have a separate wheelhouse and engine room, connected by a system of pulleys and signal bells like the Mississippi paddlewheelers of the nineteenth century. The pilots use long wooden poles to maneuver them among the flotsam of logs and swimming children; they run up and down the narrow gunwales with the lanky grace of gandy dancers.

Two months from now, in mid-October, the Juruá will be transformed into a muscular mulatto torrent that laps at the town's central square, bringing drifting logs and whole islands of floating vegetation. The flood arrives fitfully, pulsing and retreating before settling in, having less to do with local rainfall than with storms hundreds of kilometers upstream. Barges come up from Manaus during the *inverno*—winter—carrying tractors, canisters of butane, lumber, crates of canned goods. Before the first barge arrives, the whole town is uneasy and expectant because staples are in short supply. Everyone must make do. Regatões, arriving in the fast aluminum canoes used to smuggle drugs, tennis shoes, and money over the porous border with Peru, bring reports of the river's condition, and rumors flit through the market stalls as quickly as a rain squall. The Cruzeirans refer to the Juruá in muted, respectful tones as if it were a person with its own personality and intelligence. "*O Rio* fell yesterday. The barge is stuck is the mud below Boca de Lagoinha." They try to divine its next move: "*O Rio* will bring the fever this week."

The plagues arrive in early winter just as the rains begin. For a few weeks the Juruá runs foamy and roiled, washing leaf litter and logs from its banks. The water is undrinkable, with feces, garbage, and animal corpses from the bairro flutuante suspended in it. Then the river seeps over its trace, forming ephemeral ponds that linger until May and breed mosquitoes. This is the time of decline, when the old-timers wonder if they'll survive another season.

When malaria arrived in Cruzeiro do Sul during the early winter of 1933, a third of the population died in a few weeks. Almost all were elderly people and children. The epidemic was of cerebral falciparum malaria, a periodic fever that induces coma a day or two after the first symptoms appear and snuffs the babies' lives without the usual shrieking and protests of other childhood maladies. The victims never wake up. Another bout of cerebral malaria arrived in 1967 and lasted for three years. Afterward the town acquired the uncivil demography that is so characteristic of the Amazonian interior: a handful of elderly, a few children, and a shifting population of middle-aged roustabouts, fortune seekers, rubber tappers, and soldiers.

Cruzeiro's cemetery, atop the highest hill, is shaded by drooping dragon trees and pocked with rat holes. It is not well maintained. Most of the graves are marked only with simple wooden crosses painted turquoise, pink, or blue. Most of the dead are children. Some of the crosses have photos etched in porcelain, terrible images of frozen childhood.

I once walked through the cemetery with Tarzan, following the gravedigger as he turned the soil with a spade, closing the rat holes.

Tarzan asked him, "Found any bones, *moreno?*"

"Bones?" The gravedigger grimaced, resting on the handle of his shovel. "No, bones last only a few years here. Just teeth."

Tarzan, Arito, and I have been working on the waterfront for a week, outfitting our canoe for a two-month expedition. The *Fe em Deus*—"Faith in God"—is about ten meters long, crafted from wine-colored planks of black laurel, and she is as exquisitely adapted to her environment as any fish or otter. *Fe's* prow and keel—the sources of her strength—were planed from a single log, and each rib was hand-hewn from the arched wood of a buttress, conforming to the broad belly contours that give her a lazy confidence when she negotiates tight meanders, lolling and then righting herself. *Fe* has only a few centimeters of draft, the better to slide over sandbars and fallen logs. Open-sided, she is painted turquoise with green trim. Her roof is thatched with shaggy *buriti* leaves, on top of which are an herb garden planted in an old canoe and a chicken coop.

On every expedition that Arito and I have made into the forests of Acre, Tarzan has been the camp manager. A man of true liberty, he can leave town on a moment's notice. He snuffs the lantern, closes up his shop, folds his shotgun into his hammock, stuffs them both into a gunnysack, and tosses it all into the canoe. The other core expedition members are Pimentel, Zé Brejo, Edinho, and Dionesio—all veterans of botanical exploration. These men are Caboclos, smart and hard-working, and they understand the forest and camp life. Tarzan's wife, Geni, hearing tales of philandering on the river, has reluctantly decided to join us and has brought her red porcelain stove and a chair, which she places on the flattest and most stable part of the deck, where she will remain for the next two months.

Pimentel is a river pilot. An employee of Brazil's National Institute of Agricultural Reform, he makes his living shuttling back and forth to the rubber tappers' barracks, collecting tithes and transporting official travelers, bureaucrats, and the sick. He knows every meander and snag of the Juruá and her tributaries. But his is a transient knowledge, subject to constant revision. Every time he journeys on the rivers, conditions have changed: a sandbar shifts or a severed tree bobs in the current; sometimes a whole embankment slides into the water.

Dionesio, the other essential member of the expedition, is a *mateiro,* a woodsman, who can identify almost every flowering plant we find to genus if not species. Only a handful of people in all of Amazonia have this skill. Dionesio is a Caboclo of Portuguese ancestry who learned about the forest at his father's side: how to cut rubber, collect Brazil nuts, read the spoor of animals, and hunt. His father also taught him the indigenous names of many plant species in the forest, a gift of such rarity and value that it brought Dionesio his first job, with Brazil's National Amazon Research Institute in Manaus, as a plant collector on expeditions throughout Amazonia. Between expeditions Dionesio sorted specimens in the institute's herbarium, learning to translate his father's nomenclature into the Latin binomials of Western taxonomy. Now he has become an interpreter of the native and Western taxonomies.

Our farthest point on this trip will be the Serra Divisor, a range of low mountains on the border between Brazil and Peru. As the crow flies, the Serra is only about 100 kilometers from Cruzeiro do Sul, but our journey will take us through the Andean piedmont of the upper Juruá, its tributaries, and subtributaries. In all, we will thread through 400 kilometers of meanders in which the rivers cross their own path and tie off oxbow lakes, and we will have to wrestle *Fe* over hundreds of beached logs and sandbanks.

The Serra Divisor, the first substantial rise in the topography of western lowland Amazonia, consists of the eroded nubbins of a much higher range that appeared about five million years ago, when the Andes folded and uplifted. Before that time the Amazon River flowed west to the Pacific. The Serra, completely forested and rising only 200 meters above the lowland, would appear to be an insubstantial barrier to living things, but it may have been a formidable obstacle early on, rising suddenly and sharply from the plain and isolating sibling populations on either side. Mountains, even low ones, are always interesting to biologists, because they are mechanisms of isolation that create new species. We believe that the pattern of sibling species persists today, even though the Serra no longer poses much of a barrier. Other teams have collected plants on the Peruvian side; our objective is to collect on the Brazilian side. We will, as well, carry out two forest inventories in places where we will mark and identify every tree. In the years ahead we will return to see which trees have lived and which have died and how rapidly the forest is changing.

Most of our journey will be on the principal tributary of the upper Juruá, the Rio Moa, which flows through the homelands of the Poyanara, Nokini, and Náua tribes. These people have become Caboclos now and are losing their traditions, language, and culture. The Náua, celebrated for their ferocity and stubborn armed resistance to conquest by the Europeans, are commemorated in literature, art, and even a brand name of *guaraná,* Brazil's national soft drink. But the last report of a full-blooded Náua was published in 1911, a newspaper article about her marriage to a Caboclo. These days the majority of the people on the Rio Moa are more recent arrivals: rubber tappers, cattle ranchers, swidden—slash and burn—farmers.

Over the next three months, we will bivouac at the rubber-tapping outposts of Barracão Aurora, the abandoned homestead of Bom Sossego (two days upstream on the Rio Azul, a tributary of the Moa), and the Native American sanctuary of República. We will sail past the barren cattle ranch of Gibraltar and eventually reach the pioneer outpost of Fazenda Arizona at the base of the Serra Divisor. During the dry season it's impossible to go much farther than Arizona. The Rio Moa narrows and is snarled with fallen tree trunks, and except for the occasional smuggler's canoe laden with Peruvian cocaine, nobody bothers to go there. This is where the land folds and creates the first ascending waterfalls. The falls don't amount to much in the wet season—they're more like rapids—but in the dry season they are formidable barriers and you can hear their roar a kilometer away. A half day upstream from Arizona, beyond the Serra Divisor, the land starts to descend; the rivers crash down a matching set of rapids and waterfalls, and flow west into Peru. This is the watershed of the Ucayalí, the westernmost Amazonian tributary.

Fazenda Arizona is a little cell of humanity in the unrelenting wilderness. If one were to fly high above this forest at night, the home fires of Arizona—the only ones visible for a dozen kilometers—would show as a mere mote in the darkness, as insignificant as a firefly.

Right now the botanical research is a distant goal. These days are for planning. Every need for our expedition has to be anticipated, every supply purchased and loaded onto *Fe em Deus*. For the past week we have been sleeping in the canoe, guarding it as it nestles among the boats of trade. Around us, vessels drift in on the current, resupply, then move on: there are regatões, a floating bordello, a small circus, a *garota maravilhosa* (a bikinied woman who is transformed by mirrors into a gorilla), malaria control agents. Last night a preacher tied up to our canoe, perhaps attracted by her name. He roamed the beach carrying a lantern, casting light and his troubling message on the campers and sinners. The preacher left at dawn, but a rat jumped from his ship to ours and has taken up residence in the thatched roof. It is a considerate, discreet rat with no fear of humans. Good company, I think, for the trip. Lashed to our other side is a boat full of hogs and chickens, which reek of feces. A hound, enthusiastically gnawing the fleas on his shin, is chained to the prow. On deck an olive-skinned girl leans against a strut; her empty hammock, blue and white, is folded above the sleeping pigs. Her mother, big-bellied with child, is silhouetted in a window, fanning herself with a leaf. Their canoe and ours shift as any of us—including the pigs—walk around; the dog growls every time.

The first step of an expedition is list-making. Tarzan is expert at this. Here is his list of our supplies:

SUPPLIES

(For sixteen persons for eight weeks)

300 liters of gasoline, 3 ten-liter tins of kerosene, 1,860,000 cruzeiros in cash, 1 Briggs and Stratton *motor do rabo do rato,* 1 tool kit, 2 boxes of Puro Bahiano cigars, 10 spark plugs, 2 Coleman lanterns, 3 aluminum cooking pots, 1 aluminum frying pan, 96 rolls of toilet paper, 2 boxes of Omo, 5 bars blue laundry soap, 15 liters of cooking oil, 1 bucket, 30 kilos of dried beans, 60 kilos of rice, 12 kilos macaroni, 45 boxes of dried soup, 5 kilos of sugar, 25 bags of Tang, 3 kilos of garlic, 15 kilos of onions, metal plates, spoons and forks for 16, 6 packs of *anato,* 12 tins of Nescau, 21 tins of Ninho powdered milk, 10 kilos of ground coffee, 8 liters of ethanol, 15 tins of sweetened condensed milk, 5 kilos of salt, 15 400-gram tins of oatmeal, 10 cases of tinned sardines (48 tins per case), 18 bottles of tomato sauce, 18 tins of cream, 3 cases of Bordon corned beef (24 tins per case), 12 boxes of toothpicks, 6 boxes of matches, 8 cases of tinned vegetables, 12 salamis, 12 one-kilo tinned hams, 36 liters of cachaça, 1 four-liter thermos jug, 4 boxes Bom Bril, 3 wheels of cheese, 1 file, 1 grinding stone, 12 tins of butter, 16 canisters of compressed butane, 60 large plastic sacks, 50 kilos *farinha,* 6 cases of flashlight batteries (24 per case), 10 pocket mirrors, several dozen assorted Tupperware vessels with sealing lids, 20 tins of *goiabada,* 18 Bic Clic lighters, 3 shotguns, 12 cases of shotgun shells (for collecting voucher specimens), 1 box of shotgun shell casings, 1 kilo of gunpowder, 3 boxes of shotgun shell tampings, 20 kilos of lead shot, 200 tin corrugates, 200 cardboard panels, 2 plant-drying boxes, 6 Jacaré butane stoves (4 for drying plants; 2 for cooking), 2 casting nets, 200 fishhooks of various sizes, barbed and unbarbed, 12 spools 3-kilo-test monofilament, 1 *gaponga,* 6 machetes, 2 sets of aluminum pole pruners (extendible to 6 meters), 3 plant presses, 150 kilos of stacked newsprint, bound in plastic bags, 4 16-meter plastic tarpaulins, 3 kilometers of cotton string, 100 meters of nylon cord, a hammock, blanket, and mosquito net for each person, 1 backpack, 7 folding pocket knives, 7 sighting compasses, 16 1-liter canteens, 7 plastic rain hoods,

6 notebooks, 40 pencils, 2,000 numbered aluminum tags, 2,500 aluminum nails.

Supplies: where is it all stored?

MEDICAL KIT

The Merck Manual, Onde Nao Há Médico, 100 tablets of mefloquine (for malaria), 24 tablets of metronizole (for hookworms), 24 tablets of thorazine, 3 sets of surgical needles and sutures, 3 boxes of butterfly bandages, 60 500-mg tablets of Cipro, 60 500-mg tablets of amoxycillin, 60 500-mg tablets of penicillin, 50 500-mg tablets of Ceflexin, 60 500-mg tablets of tetracycline, 3 treatments of Flagyl, 10 bottles SPF-45 sunscreen, 10 bottles of Autan repellent (DEET), 20 boxes of Kaopectate tablets, 3 Episticks, 1 25-cc vial of epinephrine, 12 disposable syringes, 6 packs of gauze, 6 boxes of Band-Aids, 3 rolls of surgical tape, 2 pairs of tweezers, 2 scalpels, 2 bottles of tincture of iodine, 2 bottles of hydrogen peroxide, 24 rehydrating packets, 64 condoms, 24 bottles of chlorine tablets, 1 bottle of Mercurochrome, 60 50-mg. Benadryl tablets, 2 bottles Caladryl lotion, 6 tubes of Tinactin cream, 2 thermometers, 20 4-cc vials Butantão polyvalent antivenin for fer-de-lance and bushmaster bites.

Our last hours in town. All week we've been getting organized, tweaking the supplies, checking on details. The process seems to have no pattern, yet finally, invisibly, everything comes together and we depart. It is too late to make camp before dark, but we leave anyway, if only to watch the sun set over our destination and to sleep away from the hubbub and stench of our own species.

The mouth of the Rio Moa is only ten kilometers upriver from Cruzeiro do Sul, but once we enter it, the wilderness begins. Files of snowy egrets are sewing the gray water onto the dark sky. They roost in a grove of spiny *jauarí* palms and *munguba* trees; from afar the birds look like giant magnolia blossoms. The silhouette of the forest in the waning light reveals the varying architectures of trees and the ragged tops of bamboos. The bamboo is in flower—according to Tarzan, that is an omen of bad luck. After dark the *vagalumi* click beetles rise from the várzea side of the river among the cecropias. Each beetle has two bright patches on its thorax that shine through the wisps of vapor like miniature headlights. The constellations seem so close that they appear to touch their reflections: the Southern Cross; to its left Alpha and Beta Centauri, then Scorpio. The Big Dipper hangs just above the northern horizon, snagged on the limb of a samaúma.

We sleep in an abandoned house at Humaitá, a settlement of *crentes*, Protestant fundamentalists. The village is empty, the houses shuttered; a baleful dog is locked in one of them. Everyone in this society of zealots is wailing and chanting in the church, invisible behind a grove of mango and banana trees. Our house has a colony of leaf-nosed bats living under the floor. They are busy all night, flapping and squeaking almost inaudibly through the broken floorboards and leaving a faint scent of urine around my mosquito net. Good providers, they are bringing masticated insects and warm bat milk to their babies, which hang upside down from the floor joists. Just before dawn, the babies gurgle and squeak in response to their mothers' arrival, and then, in the near silence, I can hear them suckling.

2

RIVER OF LIGHT

I was entranced when I saw [from orbit] . . . the mirror of
the Amazon Basin, with its swamps and backwaters, like the
bewitching eye of the continent . . .

—Oleg Markarov, cosmonaut

I AM SITTING on a biscuit tin on the shaded mid-deck of *Fe em
Deus,* a day and a half out of Cruzeiro do Sul. The hens are mo-
rosely clucking in their rooftop coop; the rooster gave up greeting
the dawn several hours ago and is now asleep. Pimentel is in the
poop steering the motor do rabo do rato, his feet firmly on the star-
board and port gunwales. The helm is nothing more than a piece
of pipe welded to the chassis of the Briggs and Stratton diesel.
Squeezing the pipe between his legs, Pimentel steers the canoe by
shifting from one foot to the other while tweaking the carburetor,
getting its voice just right. The motor has no muffler, and its sharp
barks reverberate off the forest edge.

Tarzan is making shotgun shells, filling the brass casings with
powder, tamping, and lead shot, then gently tapping the firing pin
into the base. He eyeballs the amount of gunpowder, rather than
measuring precisely. This is bad news: the gun's performance will be

unpredictable. Geni sits stolidly by her red stove, boiling water for coffee and knitting. Edinho is searching among the crates for the sugar; nobody remembers where it was stowed. Zé Brejo is sharpening fishhooks with a metal file. Dionesio is reading a comic book. Arito, cigarette in mouth, has fallen asleep atop a sack of beans.

We drift dreamily through regions ever more remote. I fall into a reverie, absorbed in the panoply of the river margin. It seems like a journey to nowhere because of all the switchbacks. The sun shifts direction, first burning our faces, then our necks, as we ply the tight meanders. The river dilates and narrows, its strength waning and waxing as it slides over strange terrain eroded from the Andes. The warm surface water wanders, but the cold bottom water takes the shortest route, deep in the channel. We dangle our feet in the river as we take a shortcut across a meander and over the snags; the water feels warm, then chilly.

On the swift side of the river, a grove of dead bamboo seems to be sambaing in the current. A juvenile anaconda, yellow with black spots, suns itself on a shelf of sticks; a tabanid fly sucks blood from between the scales on its head. A family of silvery marmosets, skittering along a bough, their pendent tails drooping, look back now and then in alarm. The dominant male, about the size of a rat, bravely pauses and curses us with a staccato chipping. A morpho butterfly, the insides of its wings stained-glass blue, is intercepted by a kiskadee, which hies to a low overhanging branch and pummels it to death; the butterfly's wings adorn the turning brown water like shards of a prismed mirror. A small noctuid moth lifts off a log and enters the dangerous void over the water, where it is immediately set upon by a careening white-winged swallow. The swallow changes course, zags, dips, even flies vertically, but the moth anticipates each move. How can this moth—lungless, cold-blooded, with a brain that is no more than a smudge of neurons—outsmart and outmaneuver the sophisticated bird, an animal that knows how to navigate across whole continents and seas?

Most of the birds crossing the river are fast-moving and distant, but I can usually identify them from their patterns of flight. An Amazon kingfisher, patrolling low over the river on sharp wing, flushes convulsions of anchovies. Each time it beats its wings, the fish erupt in another spasm. A black-collared hawk has a floppy, almost gangly, flight. A yellow-ridged toucan beats its wings once or twice, then stalls and strokes again; its arched bill (which is filled with vesicles and is feather-light) creates aerodynamic drag. A black-tailed trogon is also unmistakable: a wing beat, a long, dipping glide, another beat. Three turkey vultures higher up are turning, tasting the thermals that rise from the forest. It is their season of bounty, because the river has receded from the ponds and oxbow lakes, stranding millions of fish. In this way the river nourishes the forest.

Harmony of life

Every bend in the river is the territory of a different snail kite. The Rio Moa, where moon snails as big as golf balls slide through the shallows on soft feet, is ideal kite habitat. The snails are easy pickings during the season of low water, when they lay crusty pink egg cases on the tallest culms of grass just above the reach of the hungry river. I watch a kite pluck an expectant snail from the grass, grasp her shell in his left talon and fly to a bare branch. His hooked bill severs her columella muscle, then snags her foot and entrails in one deft movement. The vivisection takes but a few seconds. The kite's bill, sculpted to the exact geometry of the snail's shell, is the snail's nemesis. The kite's bill and the moon snail are one.

Along the river's edge, the munguba trees heft hemispheric white flowers as big as soccer balls, each one with hundreds of yellow-headed stamens bent with pollen, designed to dust the night-flying bats, which will carry its germ along the river course. Each blossom is suspended over the river away from the obstructing tangle of leaves and twigs; this is a concession to the bats, which fear entanglements. The nearly unlimited light on the sunny river margin allows the mungubas to grow fast. Even their bark is photosynthetic, snatching every available photon. The light-hungry cambium peeks through the older corrugated bark like striations of green magma. The trees are in a race against time to harness the sun's energy and produce as many seeds as possible before the river reclaims its bank.

Floral scent lines, as fleeting as a breath, invisibly mingle in the stilled air above the river. These rivulets of aroma are proffered by orchids to lure trap-lining bees and wasps. Now I smell the musk of a family of long-nosed bats, the color and texture of peeling bark, which are sleeping under an overhanging log. I can't discern the pattern of fur and limb until the last minute when, in an instant, they scatter like a flurry of brown leaves.

Opposite the deep-cut riverbank are heaps of sun-white sand fringed by willows. The beaches are etched with signs of life: the Escher-like footprints of a troop of coatimundis, the folded trail of an anaconda, the long-toed tracings of a tegu lizard (bisected by the mark of its steady tail), the wispy five-fingered touch of a heron. An *Ameiva* lizard sprints across the hot sand from one shrub's shade to another's, then waits stick-still for a beetle or fly. A demolished peccary—a bonanza of salts and minerals—is buried in a flurry of yellow butterflies, each dipping its coiled proboscis into the decomposing stew.

On the sandbank ahead is the spoor of a caiman. Judging by the splay of its four-toed footprints, it was easily four meters long.

There are telltale signs of struggle: deep prints that bespeak exertion and characteristic luffs of sand showing that the caiman was rolling with the prey in its mouth. Nearby is the severed head of an electric eel, a *poraqué*. We pull ashore to investigate. Careful: the caiman may still be close. The eel's lozenge-shaped head is as big as a melon; its body must have been at least two meters long. It is fresh and free of flies; this battle just ended. I can discern the sensory crypts that detect subtle distortions of the eel's own electrical field, giving it a preternatural ability to detect prey in the opaque river. Its nearly vestigial eyes are fogged with cataracts—a common condition in electric eels—probably the result of repeated shocks.

I don't understand the circumstances of this battle. Why would a caiman attack an animal that can generate 500 volts—sufficient to stun it into unconsciousness—and then, enduring the torment of electric shock in its jaws and brain, drag it to shore, roll, and flail it to death? There must be far easier prey in this river. But Arito, who was once a commercial caiman hunter, understands.

"It was an accident," he explains. "Caimans are stupid. Their instinct is to clench their jaws when they get into trouble. They probe the muddy river bottom with their huge mouths and simply clamp down on anything that moves: fish, turtles, other caimans—and eels. All this caiman could do when it felt the shocks of the eel was to convulse and bite tighter, to tough it out until the eel died. It wasn't bright enough to let go."

Arito laughs. "It must have fried its brain."

We anchor for lunch in the shade of a spreading *fava de tambaqui* tree overgrown by a *flor de mulher* vine, which has dropped purple and white petals into the still water. The vine above is barren, but its reflection is in bloom. We are in an eddy, on the edge of a green meadow of floating grass, ferns, bladderworts, and lily pads, and the whole terrain undulates in the wake of *Fe em Deus* as as we tie up. A wattled jacana delicately splays its long toes as it walks across the lily pads without tipping a single floating leaf. It is a male; the crooked toes of his chicks dangle from beneath his wings like a bundle of sticks.

The water surface, the boundary where the air teases the water, is the most extensive biome on Earth. Here the skin of the river is reinforced by organic matter, a film of pollen, oils, and detritus that varies in thickness depending on the amount of local decomposition and the muscle of the current. Today the scum is heavy; the eddies that spin off our paddles dent the water. Whole communities of plants and animals, rafts of logs and vegetation accumulate in these stilled waters and wait for the river to rise. Most of the vegetation consists of *Pistia,* an aquatic relative of the philodendrons that grows in rosettes, trailing its yellow roots in the river, and *capim,* a grass with buoyant, jointed stems that spreads over the surface of the river, creating an itinerant floating canopy a few centimeters tall. A nomad, capim lives in the corners of the river, moving on when the water rises, but when stranded it can set roots into the wet mud and invest in permanence. In the lower Amazon, rafts of capim as large as small towns tear lose from the banks during the rainy season and float out to sea, bearing iguanas and even the occasional fer-de-lance.

The capim, its seed heads drooping at this time of the year, is finely hirsute; the hairs trap a layer of air that beads the rain as if it were skittering on oil. Naturally, there are grazers on this bounty: endemic grasshoppers with yellow-and-black-banded legs, crooked on the leaves of grass; moths, gnats, water striders, flies, mosquitos, and ants all prowl the capim or dance on its surface. And predators: legions of spiders, whose eyes scatter the sunlight like diamonds. Some hop on the skin of water like skipping brown stones, bounding after small prey.

Parts of the floating prairie are covered with bladderworts, which grow frilly leaves among the drifting puffs of algae, the blooming phytoplankton, and the abounding zooplankton. The carnivorous bladderworts produce hair-trigger underwater vesicles that implode when brushed by a passing copepod or larval midge, engulfing and digesting it. Above the river's surface the bladderworts' redolent yellow flowers entice enthusiastic bees in a more benign cause: to collect and deliver their pollen. The bladderwort hunts the water plankton, but it seduces the air plankton.

The long stretches of forest are broken by a few clearings—pioneer outposts. Every homestead has a dugout canoe hauled up on the embankment. Each canoe is a work of art, and each represents liberty. The homesteads seem like still lifes as we drift by. The people watch our passage, their daily rites momentarily frozen. A mother is bathing a new baby, like a crumpled pink leaf, on the beach. A young woman, her dress sliding off her brown shoulders, grinds corn in a pestle made of *pau d'arco* wood. A rooster mounts a hen, who squawks with indignation, ruffles her skirts, and runs from his advances like a coquettish maiden. A lanky boy climbs into the loft of his house to retrieve sheaves of corn, then decides to stay and take a nap. Three men are castrating a pig, two restraining his legs while the third slashes. Even the pig stops struggling to watch our passage, then runs off, screaming and bloody, into the forest.

A crone, bent like a Makonde sculpture, is manicuring her herb

anted in a broken canoe, of chives, coriander, jambu, and
s. She chucks the weeds at the chickens and tosses us a
d, which is used to scent corpses in Amazonia.
māe!" Tarzan laughs, recoiling. "We're not dead yet!"

The sky above the river is cloudless, but over the hot land, the forest breathes vapor. The transpired water erupts into clouds of small consequence, save for the roving shade they cast over the canopy. In the dry season these clouds build up rapidly after dawn, becoming popcorn cumulus, then dissipate quickly in the late afternoon. Their energy is too centrifugal to create a column of strength, so the atmosphere's stored energy dissipates, and the nights become star-clear. But today, near the onset of the rainy season, the air is nearly saturated, unable to hold much more water. The clouds linger, sucking the breath from the land, and coalesce, eventually making the huge anviled nimbocumulus columns that puncture the cold stratosphere, creating the friction among the furious particles of air that are the forges of lightning.

Even though we are only about two hundred kilometers from the Pacific Ocean, most of the air and clouds overhead originated on the other side of the continent in the Atlantic trade winds. During its journey from the Atlantic to the Andes, water vapor is repeatedly absorbed and recycled by the vegetation. In fact, the typical Amazon forest returns roughly half of its own rainfall to the sky through evapotranspiration. Rainwater is absorbed by roots, transported in the vascular tissue of trees, shunted through the metabolic pathways of plants, utilized in various physiological processes, and eventually released back into the atmosphere through the leaves. Another quarter of the rainfall evaporates from the surfaces of trunks, branches, leaves, and other components of the vegetation. Only a

quarter of the rain is carried away by the rivers. The forest, therefore, creates about three-quarters of the moist climate on which it
depends. The forest and the air above it are an integrated system.

At the Andes, the vapor piles up, unable to go farther; then, starting in October, rain pulses eastward in rebounding waves, returning the day's vapor to the land. Thus the rainy season in Amazonia
originates in the west and spreads east, even though most of the
rainwater comes from the Atlantic. And it's rough air up there. Vultures bob on the thermals, tasting the upwellings for the scent of
decomposition, those few molecules of cadaverine that signal dinner. Today, by midafternoon massive storm clouds and black hori-
zons of rain have formed. Two scarlet macaws are dollops of red in
the green canopy; against the dark western sky, they seem radiant.

At dusk, lightning rends the turning clouds, a harbinger of the
rainy season. Downdrafts, cold and heavy with moisture, tousle the
river surface and the treetops. The air smells of fresh ozone, created
by the electrical discharges: oxygen molecules are being cleaved and
smashed together at temperatures as high as those on the surface of
the sun. This process is one of the sources of the ozone layer, which
protects the billions of leaves and scales and butterfly eyes—the
whole living skin of Amazonia—from being frazzled by the sun's
ultraviolet energy. In this manner the forest waters and guards itself:
the sun's energy drives photosynthesis, which gives trees the energy
to suck rainwater through the dark and viscous xylem of trunk and
leaf, to be released again to the air.

We are sailing through an uneasy terrain: ephemeral islands of
capim, sandbanks colonized by pale green willows and strong but
yielding bamboos that contest the river's muscle. The bamboos
are strangely uniform, all the same height and age for hundreds of

miles along the river course. Arito remembers why: in 1984 the bamboos simultaneously flowered, died, and started anew from seeds. These plants are all the same cohort. On the upper reaches of the riverbank, where the current is less angry, the first cecropias appear. Their lobed leaves are the size of Frisbees, green on top and silver on the undersides, and the strong winds that race along the river turn them in sequence, like a stadium full of placards waved by fans. Behind the cecropias are pure stands of jauarí palms, whose trunks are armed with ten thousand brown spines designed to foil climbing rats, and lovely palmate-leaved pupunha palms, which produce bunches of fruits as polychromatic as an artist's palette. Finally, as the várzea grades into the woody trunks of permanent forest, we see the pale-trunked *ucuüba* (related to nutmeg), and the buttressed samaúma trees.

Although they would never build a house on the floodplain, the Caboclos know that the várzea is a place of nourishment, a forest of survival that fosters their permanent settlements on the nearby *restingas,* or ridges. The várzea soils are renewed by the *café-au-lait* floodwaters that annually decant their Andean sediments. Every year the departing waters leave a layer of new fertilizer. On the alluvial lenses left by the receding waters the Caboclos grow beans, peanuts, and maize—a medley of crops as old as the Americas—and even a few woody plants that like to have their feet in the water: passion fruits, cacao, and coconuts. For a season, the river margin becomes farmland. Although fertile, the várzea is dangerous for the farmers. Timing is everything. The Caboclos must plant their crops at the first opportunity after the floods recede, but it is folly to start before the water levels have stabilized. To get a head start on the growing season, the settlers germinate their seeds on high land in a rolled banana leaf or an old canoe converted to a planter, allowing the striplings to grow for a month or two before being transplanted to the beach. Likewise, at the end of the growing season, they must bring in their harvest before the waters return, but not before the crops have ripened fully.

🐚

On this afternoon just before the onset of the rains, the fish are hunkered down in the belly of the river, waiting for the water to rise. August is their leanest time. The low waters are crowded, food is scarce, and the fish can't swim into the nutritious flooded forest. The river has become a place of nearly unlimited protein, and the Caboclos feast. If they can get salt—which is worth its weight in gunpowder in much of Amazonia—they rub it into their surplus catch and hang the split flesh on lines to dry. Set to dry in blue-eyed rows, the ocellated *tucunaré* look like strings of prayer flags.

While the others nap, I drop a hand line over the side of the canoe. Fishing in the Amazon during the dry season provides immediate gratification. The celebrated fisherman's patience is not required. My hook is struck immediately by a small *piranha preta,* too skinny to keep. Within thirty minutes, I've hooked as many fish: ten *piranhas caju,* two *traíras,* two *aracus-pinima,* a dozen silvery *pacus,* three *botinhos,* and a small *tambaqui.*

Each of these species of fish feeds on different things, and you can tell them apart by the way they bump the bait—a sort of taxonomy of nibbles. The predatory piranha caju bites hard and tugs the line downward. The piranha preta, which has flat, serrated teeth for stripping the flesh off fruits, rolls the bait in its lips before biting. The tambaqui has seed-crushing back molars as flat and wide as a human's, which can demolish the woody jauarí palm fruits that I can split only with a machete. Its bite is as strong as a vise. The tiger-striped aracu-pinima, which has teeth that pick insect larvae and algae from logs with the precision of tweezers, is tentative, ever probing with its pointed snout. The tucunaré, which swallows fish whole, hits the bait hard and runs with it. The pacu, whose teeth are designed to shear leaves, nervously fumbles it. The botinho, a bottom feeder that probes the sediments with sensitive, tactile barbels, steals the bait; you hardly sense that the fish has touched the hook.

Sometimes the tambaqui will wait under a gravid rubber tree, listening for the splashes made by dehiscing fruit; likewise, the black piranhas congregate under the ripe mungubas, waiting for the fruits to drop. To catch them, the Caboclos use a gaponga—a stringed cork that they plop into the still pools, an acoustical bait that mimics the sound of the falling fruit—or snap their fingers under the water.

The fish may be hungry and easy to catch, but they are hardly satisfying to eat. The piranhas seem to be all ribs. The tambaqui are hollow-bellied and lack the sweet translucent calipee they have at the time of flood. And they are gorging on the bad-tasting *fava bulacha* beans, whose aroma is evidently attractive to fish but is repulsive to humans. When the tambaqui eat them, their flesh acquires the same stench.

 We will eat corned beef tonight.

Today, in late September, the river and land are expectant. In the course of the next month the rains will gradually increase in both frequency and intensity. The shift of the várzea forest from terrestrial to aquatic is one of the great seasonal changes on Earth, visible even from space. The river rises in pulses. The early floods, born less than a hundred kilometers west of here, last only a few hours, but with every pulse the baseline water level rises a bit. One day the river is obstructed by sandbars; the next day it is a torrent. The next morning it is low again but higher than before. You must anchor a canoe with a long tether during this time of transition, with sufficient slack to keep it from being pulled underwater.

By February the Rio Moa will have risen ten or fifteen meters, overwhelming its banks and spreading through the trees. You can canoe through the forest then, floating from one canopy to the next, where today you walk on trails of cracked mud. Each emergent tree is an island unto itself—a refuge for thousands of spiders, scorpions, centipedes, pill bugs, beetles, and ants—and because of the random assembly of these refugees, every tree becomes an ecosystem distinct from its neighbor. In areas where the trees are widely spaced, every tree has its own boa, the bats it feeds on, and maybe an iguana; every cecropia has its own three-toed sloth. The dusky titi monkeys, which almost never come down to the ground, are specialists in this aquatic forest, and a passing troop can be a local disaster, knocking myriad animals and detritus into the water. The *aruanã* fish follow the monkey troops through the forest. Their upward-looking eyes are designed to survey the treetops, and their lips are armed with needle teeth. Swimming just below the surface, the aruanã presses its two forked barbels against the surface tension to taste the air currents and sense disturbances, the splash of a centipede perhaps. It is especially fond of monkey feces.

The fish swimming through the várzea disperse fruits and seeds just as any trogon would in the terrestrial forest. The trees have coevolved with the fish, which carry their seeds upstream; the nutritional virtues of their fruits have been specifically crafted to attract and satisfy one or several species of fish. And the fish understand the trees, know their phenologies, their economies. They grow fat on this knowledge.

Every year billions of fish belonging to hundreds of species migrate up the tributaries of the Amazon into the flooded forests. The migration, known as the *piracema,* begins in June or July and lasts until the várzea floods in December, when the fish disperse among the trees. It is an event of great urgency. Sometimes the water is warped with millions of fins. The silver-sided pacu look like pale leaves tumbling through the flooded boughs, and the air above the river fills with a pervasive buzzing, the collective raspy voices of innumerable *pescada* and *surubim* stridulating their swim bladders with a bone. Like so many earthly migrations, the piracema seems to coincide with each full moon. How, I wonder, as the fish grope though the turbid waters, feeling their way with lips and barbels, do they sense the phase of the orb above? Are they drawn, perhaps, by small tides?

The late-phase piracema, when the fish enter the forest, is a time of convergence, too, bringing hungry river turtles, giant otters, six-meter-long black caimans, freshwater dolphins, and humans together to feed in the muddy waters of the flooded forest. Fishing is harder then than in the dry season, because the fish have lots of options. They will disdain bait if it smells even slightly suspicious—if it has been contaminated, for example, by the oil of an inopportune finger. A fisherman needs patience at this time of year. But the harvest is worth the effort. The tambaquis are so plump that their fat drips onto the fire when they are roasted. The meat of the dorsal muscles of the cashew piranha is dark and gamy, even a bit bitter, but the rib meat is as yellow and sweet as butter.

In the late afternoon the air chills and the warm river breathes life. Three columns of midges, like strange animate storms, rise two hundred meters above the river, braiding and separating in the sunset. Each column contains millions of animals, each no bigger than a grain of pepper, each clearly visible against the western sun, yet each untrackable. This is probably a communal orgy, an efficient way for these minuscule animals with a lifespan of a few hours, living in a spatially complex environment, to find mates. It is a vortex of sex, the midges coupling on the wing, tumbling in pairs through the air among their uncountable brethren, all similarly united. Trillions of sperms are being ejaculated overhead.

Pimentel cuts the motor, allowing the canoe to drift under one of the columns. As if it has a collective mind, the column lifts its skirts, allowing us to pass beneath. The tornadic air above is darkened with midge bodies, creating a thrumming song that has no particular origin, like the sound of a distant tuning fork. The air in the wake of the column is cloying, suffocating, depleted of oxygen, and slightly warm; it takes a lot of energy for forty million wings to loft a ton of midges a quarter of a kilometer into the sky.

After the sun descends, thick and red, its light skips across the undersides of the clouds, illuminating only their inverted ridges. Across the river every tree is etched against the horizon, as sharp as cut paper. The last light is striated by the saw-toothed Andes, unseen to the west.

We camp on a sandbar next to the terra firme forest. The woods are murmuring, shouting, screaming, croaking, rustling with crawling things that can see in the night. I can't differentiate the individuals in this vast chorus of insects and frogs. I lose track when the count reaches ten or twelve. One tree frog in particular is especially loud, its call a graceless *araouk, araouk*. The Caboclos call it the *canoeiro* because its voice sounds like the scraping noise (and pace, too) of someone slowly bailing a canoe with a gourd. A potential mate (or is it a rival?) *araouk*s across the river; its call is always out of phase with the first one's. The nighttime voices have to be rigidly partitioned to avoid interference. Different voices of similar timbre never compete; one always listens when the other is active. Is there an economy of voices? How many individuals (or species) can share a particular frequency, a specific pitch, before that acoustical niche is used up? What about the ultrasonic frequencies that I can't hear, used by the bats? Are they partitioned, too?

A new moon like a bruised eye, a few hot stars. A *rasga mortalha* flies overhead, calling *chica, chica, chica . . . ghrrrr* in the darkness. The Caboclos believe that it is a soothsayer, that someone will die soon.

3

AN EXPEDITION OF POETS

Toward what inevitable new disasters
are lured these poets, these peons?
To what perils, what misery, what destiny
in service to some high-born name,
What celebrity have they been promised? What fame?
The spells of kings; of veins
Of gold, made so casually, so easily?
What triumphs? What applause? What victory?

—Luis de Camões, "The Disenchantment of Eldorado"

A DAY LATER, at noon, we reach a wide bend of the Rio Moa where the forest briefly parts and the sky, reflected on the dark water, makes an inverted horizon. Nestled in a grove of spindly *açaí* palms on a high bluff of terra firme is the rubber tappers' settlement of Seringal Aurora, an outpost of about a hundred men, a few women and children, a dozen pigs, and innumerable chickens. Aurora has a small supply store and a *barracão* where the unwed *seringueiros* (rubber tappers) live in a lonely and boozy society, ungraced by domestic pleasures. We will stop for the night and spread the word that we plan to hire ten good workers, at decent *diárias,* for the next eight weeks. The latex is running poorly this year, and we can be choosy: nobody with jaundiced eyes, too many obvious knife wounds, or booze on his breath. Each man has to bring his own canoe.

The dormitory floor, made of splintered trunks of buriti, is perched on stilts above the pig pen. We hang our hammocks on its sagging verandah. I write at a long bar crudely hewn from the local *cedro* trees, sipping a series of unsatisfyingly hot beers. The pigs snort and splash below. Overhead on a rafter, an *acaba da igreja* is plastering pellets of masticated mud onto her wattle nest; she flies to and from the riverbank, fetching more mud.

The primary pastime here seems to be waiting—for the next canoe, for the regatão, for the river to rise. The heat dilates every activity. Two boys are slouched like ebony sculptures across the threshold. Inside, all six hammocks are pregnant with indolent men. The breeze shuffles the palm leaves and scatters the big white butterflies that are sipping nectar and fornicating in the weeds, but does not stir the heat.

The mateiro, or manager of the rubber estate, the *seringal,* is Chico. His house, painted pink, shaded by açaís, its windowsills accented with green, is across a clearing. Chico's thick-legged pregnant wife wears only a short nightie. She is boiling manioc, the bland belly-food of this impoverished forest, and making *oaca* pellets, a mixture of flour, buriti fruit, and red pepper. When fed to fish, oaca causes them to bloat and bob to the surface. A naked child waddles everywhere, pacifier in mouth, excreting indiscriminately. Chico, bored with bookkeeping, anxious to go fishing or hunting, stumbles around the house and yard, pauses under the spreading *Terminalia* tree and finally retires to his hammock, stationed strategically by the door. Now the black and white dog who lives under the bar dashes across the shaded balcony and barks at phantoms in the forest. A heroic one-word vocabulary. His adoring bitch, tan and white, has swollen nipples and a severe case of fleas; she watches with a crooked eye.

Too sleepy to write now, I lie in my hammock and listen to the shuffling leaves of the palms, the rustling of the river, the distant shrieking of a pair of scarlet macaws, the creaking of a tree frog in the barrel of rainwater behind the stove. But most especially I listen to the voices of the *japiims,* whose nests hang in the boughs of the samaúma tree that shades the barracão. The japiims mimic the voices of eight or ten other birds—from parrots to toucans to wrens—making their colonies sound like a mixed feeding flock. They can mimic mammalian sounds, such as the screams of otters and human snoring, and even waterfalls. A colony in full voice produces an ever-varied melody reminiscent of water tumbling over rocks, a sound as soothing as rain.

The japiim nests—several dozen of them in a group—are woven from pliable grass and hang like round-bottomed amphoras, the eggs and chicks safe inside deep, soft pockets. Not just any tree will do. Although they are indifferent to the humans below, the japiims choose their other neighbors carefully. Somewhere near each colony—often right in the middle—is a nest of predatory wasps, which busily bring insects, paralyzed by their stings, to feed the larvae snuggled in the nest's hexagonal paper cells. The japiims fly through the swarming wasps, which tumble in the vortices made by the birds' wings but do not attack.

Why would these birds, which could weave their nests on any bough in the forest, choose to raise their babies next to hundreds of voracious, meat-eating wasps? The answer is that the wasps protect them from botfly maggots, which infest the flesh of warm-blooded animals. Botflies have been clocked flying several hundred kilometers an hour, perhaps the fastest of any animal on earth. In the blink of an eye, a female botfly can dart into a japiim nest, lay her micro-

scopic egg on a chick, and escape before the parent notices anything. Soon the egg hatches into a ravenous maggot that will burrow into the chick and eat it alive from the inside out. The wasps,
however, which also live by the microsecond, intercept the flies on
the wing, paralyzing and feeding them to their own hungry larvae.
To the wasps the japiims are slow-moving lures that attract the botflies. To the japiims the wasps are near-supersonic rapiers that guard
their babies from enemies as invisible as the air itself.

Each of the seringueiros at the Seringal Aurora is allocated one or
two *estradas,* sinuous trails that thread through the várzea and terra
firme from one rubber tree to the next. An estrada is usually four
or five kilometers long and links eighty to a hundred trees. In exchange for the privilege of tapping these trees, the seringueiros are
indentured by debt to the mateiro, the surrogate manager for the
patrão, or landlord, who lives in Cruzeiro do Sul. The mateiro is the
absolute ruler here, both boss and banker, allocating estradas and
selling each seringueiro an *aviação:* a grubstake of coffee, salt, sugar,
gunpowder, a vial of penicillin, a bottle or two of cachaça, a kilo of
lead shot, batteries, and the two hundred or so tin cups (usually
used corned beef tins) in which to collect latex. The tappers are
contractually obliged to sell their rubber to the mateiro in exchange
for the essentials of life on the frontier. He almost never pays in
money, whose value is not clear on this frontier and rapidly becomes useless through inflation. You can get cheated of your money
or robbed. Besides, the mateiro knows that cash is a seringueiro's
ticket to town, and without it, he must stay. The barter system traps
the serengueiro in the wilderness.

Rubber tapping is an exhausting profession that takes up most
of the day and night. At night, when the sap runs most copiously,

the seringueiro must patrol each estrada twice: first to score the tree and attach the cup, and a few hours later to collect the latex. During the day, back at camp, he must slowly dribble the latex over a smoky rotisserie so that it coagulates into a bola. No single human can sustain this pace night and day for very long. To survive, a seringueiro must take a wife who can perform the daytime chores while he sleeps. The couple builds a little homestead close to their estrada: an open-walled house on a platform of split buriti, a few posts from which to suspend hammocks, and a thatched roof. Since seringueiros begin their working lives just after puberty, they necessarily marry young, and the generations are only fourteen or fifteen years apart. It is common in the Amazon to see men in their fifties cutting rubber with their great-grandsons. If the region had good medical care, the early onset of childbearing would create a steep population growth rate, but on the Rio Moa, where so many children die in infancy, the numbers of seringueiros are dwindling.

There are, as well, a few unmarried women on this frontier. Two, María and Manduca, have lived near Seringal Aurora for thirty years. They have a flock of kids of a variety of colors and physiognomies, whom they raise as one family. Some of the children were adopted, others were sired by regatões and seringueiros who passed by over the years. The men take pride in these children and bring them gifts and supplies. When Pimentel and Tarzan drop off sacks of flour, María runs down the beach after them, bidding us welcome. Geni looks at her askance and orders us to move on.

Once thousands of rubber-tapping outposts lined the tributaries of the western Amazon. Now only a handful remain, survivors of an era of discovery, boom, and bust, of the extinction of native peoples and the enslavement of thousands of new arrivals. The story of rub-

ber begins in April 1736, when a ship dispatched by the French
Académie Royale des Sciences in Paris, on the order of King Louis
XV, arrived in the disheveled port city of Esmeraldas on the coast
of Peru (now Ecuador). The purpose of the expedition was geographical: to measure the angle of one degree of meridian against a
fixed star simultaneously from two separate locations on the equator. (The expedition was complemented by a similar one to the
high latitudes of the Arctic; by comparing the angles, one could
calculate the sphericity of Earth.) The expedition leaders, Charles
Marie de la Condamine and Pierre Bouguer, were in their thirties,
and they kept voluminous journals, recording every detail of their
adventures. They became the European discoverers of hundreds of
new species, including the South American rubber tree.

The two explorers joined a caravan that took them east across
the Pacific plain and up the escarpment of the Andes to the colonial capital of Quito. During that long trek their paths were lit at
night by torches of a coagulated tree sap, which la Condamine described as "black and resinous," wrapped in what he believed to be
banana leaves. He collected some of the sap and sent it from Quito
to the Académie Royale with the following description:

> In the forests of Esmeraldas province a tree grows which the natives
> of the country call hévé; simply by an incision it lets flow a white
> resin-like milk; it is collected at the foot of the tree on leaves specially
> spread out for it; it is then exposed to the sun, whereupon it hardens
> and turns brown, first outside, and then inside. Since my arrival in
> Quito I have learned that the tree which discharges this substance
> grows also along the banks of the *Amazon* river, and that the Maïnas
> Indians call it Caoutchouc; molds of earth in the shape of a bottle are
> covered with it; they break the mold when the resin has hardened;
> these bottles are lighter than if they were of glass, and are in no way
> subject to breakage.

This passage is the first written description of Amazon rubber. La Condamine heard stories that the Omáguas, once the mightiest tribe of the Upper Amazon, used the resilient product to make toys:

> The *Omáguas* make squirts or syringes thereof . . . [they are] made hollow, in the form of a pear . . . leaving a little hole at the small end to which a pipe of the same size is fitted; they are then filled with water, and by squeezing them, they have the same effect as a common squirt. This machine is mightily in vogue amongst the *Omáguas;* when they meet together by themselves for any merry-making, the master of the house never fails to present one to each of his guests; and the use of the squirt with them is always the prelude to their most solemn feasts.

La Condamine and Bouguer spent the next eight years in the highlands of Ecuador, eventually making a successful measurement of the arc of meridian in March 1743. In July of that year la Condamine began his journey back to France, but instead of going by way of the Pacific, he descended the escarpment of the Andes to Iquitos and continued down the Amazon 3,000 kilometers to Pará (now Belém), "a route," he wrote, that "no one would envy." He described an unpopulated river; the Native Americans had been reduced to vestigial numbers by disease and slavery and the survivors had dispersed deep into the forest, away from the routes of commerce.

La Condamine must have sailed past the maze of distributaries, cloaked in várzea, at the tawny mouths of the Rios Juruá and Purus. Then as now they were a places of disorientation, mosquitoes, and pestilence. He must have observed the broken logs and debris dragged from 1,000 kilometers distant by the angry rivers. These rivers, for sure, bled a vast terrain. But there was no sign of the wealth they secreted, no hint that one day the rubber trees of

Acre, the land drained by these rivers, would be coveted by the world.

Belém, where la Condamine arrived on September 14, 1743, was an outlier of an empire made wealthy by trade in cacao, an enclave of pampered plutocrats surrounded by a forest of infidels who murmured in strange tongues and who brought strange items to market. La Condamine found Belém in the throes of a smallpox epidemic that was decimating the indigenous population but to which the Europeans were for the most part immune. La Condamine was stranded, a feted captive of soirees and glittering parties, while people were dying by the thousands all around him. The epidemic made it impossible for him to continue his explorations in lower Amazonia; there was no one to paddle his canoes. La Condamine left Belém in December, shipping out for Cayenne and on to France. On April 28, 1745, he addressed a packed public assembly of the Académie Royale des Sciences, demonstrating the strangely elastic properties of caoutchouc, revelations that fascinated the industrialists of Paris. Neither the Americas nor Europe would be the same afterward.

A century and a quarter after the presentation to the Académie Royale, Amazon rubber had become all the rage in the industrialized countries. The Americans had invented vulcanization—the process of heating crude rubber with sulfur—to make a product that was not only elastic but durable and that could be used for making gaskets, tires, belts, raingear, and countless other products that no one could have imagined in la Condamine's time. Pound for pound, rubber was one of the most valuable raw materials on Earth. And the newly independent countries of Amazonia held an absolute monopoly on the substance. Inevitably, foreign powers coveted this treasure. James Drummond Hay, the British consul at Pará in the 1870s, wrote:

> The labor of extracting rubber is so small, and yet so remunerative, that it is only natural to mankind, especially to the yet sparse and comatose population of these provinces, to prefer that occupation; in which a family gang, or single man, erect a temporary hut in the forest, and, living frugally on the fruit and game which abound, and their provision of dried fish and farinha, realize in a few weeks such sums of money, in an ever-ready market, with which they are able to relapse into the much-coveted idleness.

The Yorkshire plant explorer Richard Spruce observed, "The extraordinary price reached by rubber in Pará . . . at length woke people up from lethargy . . . The mass of the population throughout the Amazon and principal tributaries put itself in motion to search out and fabricate rubber."

By the late 1800s rubber was Brazil's most valuable export commodity; Acre had the greatest expanse of rubber trees in Amazonia and the highest-quality trees as well. At first, most of this extractive economy was based on the virtual slave labor of Native Americans, enforced by a system of debt patronage and homicidal *patrões*. Soon Eurasian and African diseases—smallpox, yellow fever, measles, and malaria—arrived with every boatload from the east, and the isolated work camps became places to die. Most of the Native Americans in the rubber-tapping areas of western Brazil and Peru perished during this era. According to one estimate, half of the native population of Acre died just in the peak years of extraction from 1900 to 1913.

By necessity, the rubber barons sought other sources of labor, particularly in the impoverished northeastern state of Ceará. Periodic droughts had long afflicted the northeast, but the drought of 1877–79 (which we know now was a particularly severe El Niño year) was the worst in history. The economy of Ceará was based on two agricultural export commodities: sugar and coffee. The coastal forests had been destroyed for these monocultures and had lost the ability to generate their own rainfall. During these times thousands of farmers in the northeast had no means to feed themselves. A restless and destabilizing proletariat, they had also become a political liability. The convenient solution was to deport the impoverished masses to greener fields, and the hinterlands of the Amazon were the logical destination (this policy was repeated in the 1970s during the colonization of the Transamazonica).

Approximately a hundred thousand people—entire extended families—were transplanted from Ceará to Acre, a journey of twenty days and twenty nights, 2,400 kilometers by ship from Manaus to the mouth of the Juruá-mirim. In exchange for their passage by steamship and a modest grubstake, the immigrants were obliged to cut rubber in the wilderness estates and to sell their product only to the patrões. The 500-ton steamship *Lucania* was typical of the vessels that made the voyage. She had two salons, a dining room, and several first-class cabins, but most of the passengers—the bonded immigrants from Ceará—slept in the open air in hammocks. Euclides da Cunha, who chronicled that experience, wrote:

> They overcrowded the vessels, the canoes, the steamers with their huge human cargoes, consigned to die. They shipped us to Amazonia—vast, uninhabited—to become exiles in our own country. The martyred multitude, bereft of rights, severed of their familial ties, the shattered tumult of sudden parting, bearing a letter of employment for the unknown; they leave with the starving, the fever-wracked, those afflicted with smallpox; in conditions to make malignant and corrupt the most salubrious regions of the earth.

Unlike the California gold rush of a few decades earlier, where a single nugget could make a man rich, the Acre frontier was not a free-for-all of small claims. Crude rubber, however valuable, was a bulky product that demanded economies of scale in order to be extracted and marketed profitably. A few of the immigrants from Ceará—heavily capitalized by the emporiums of Manaus and Belém and accompanied by armies of hired pistoleiros—declared themselves masters of vast tracts of riverside forest; some of the same *seringais* persist today. Fortunes were at stake on the frontier; this was land worth dying for. The only law was, in the parlance of the times, "article 44 of the legal code." In other words, conflicts were settled with 44-caliber revolvers. Once established, the seringais adopted rigid military hierarchies. The owner, who bore the ceremonial title of "colonel," commanded various ranks of *tenentes, subtenentes, majors, sargentos, cabos,* patrões, and mateiros. At the bottom of the pecking order, of course, were the indentured seringueiros, who, as today, had to pay the colonel 40 percent of the value of the rubber they produced.

By the late 1880s, the rivers of Acre were strung with seringal fiefdoms, each under the mortal command of a colonel. One of the most powerful colonels was Arito's maternal great-grandfather, José da Silva. Born in France, José emigrated to Acre in the 1880s with his Cearense wife Mariada Luiza. He claimed a long stretch of forest at Tejo along the margin of the Rio Juruá Mirim. Soon the couple established another seringal at Iraçema, a day farther upstream from Tejo. Within decades the Seringais Tejo and Iraçema—where a single estrada could yield as much as 900 to 1,000 kilos of high-quality rubber per year—had became the most valuable real estate in Acre. At Iraçema, where for hundreds of kilometers in every direction there was nothing but forest and river, José and Luiza built a great wooden house with six rooms and a huge kitchen, all surrounded by balconies. They imported a grand piano and oak cabinetry from Portugal, crystal glasses and bone china from England.

Luiza gave birth to five boys and four girls in that isolated spot. All of her children went to the coastal city of Fortaleza to be educated in arts and letters; three of her daughters studied in Portugal.

The barracão at Iraçema housed eighty or ninety seringueiros. The da Silvas' control over their lives was absolute. When a young seringueiro asked the colonel's permission to marry his sweetheart, the colonel ordered him to marry her widowed mother instead, a woman whose advanced age made her more desperately in need of a man. The seringueiros were a demoralized, brutish lot, with little to lose and a lot to gain by challenging the colonel and his wife. Well-paid bodyguards, known as *capangas,* protected the main house day and night. Luiza, a tiny woman, just over five feet tall, carried a 38-caliber pistol at all times, even to her bath. One afternoon, when she discovered a bull eating her vegetable garden, she shot it dead square between the eyes, while the men in the barração watched respectfully. Nobody dared challenge her.

This was Indian country. Seringal Tejo was in the territory of the Kampa Indians, Iraçema on the seasonal migratory route of the Amuacas. Eventually, José and his sons managed to placate the Kampas, but the Amuacas never submitted, often attacking and laying siege to the seringais. The colonel regarded the Indians as impediments to modernization and progress—not to mention profits— and ordered that they be hunted like game animals. He employed squadrons of professional hunters, known as *batedores,* to steal along the rivers through the forest and observe the local tribal settlements, waiting for an opportune time to attack. These hunting parties, or *coreiras,* often took advantage of religious festivals known as *caiçumas* or *mariris,* when the Indians drank fermented manioc, known as *chicha,* and a hallucinogenic decoction made from *caapi,* the *Banisteriopsis* vine. The batedores waited until the Indians were delusional, then sneaked into the camp, cut the strings on the Indians' bows, and threw the arrows into the fire. Retreating to the forest, they gunned down the disoriented and unarmed warriors with their Winchesters. The coreiras were purely economic campaigns and therefore were calculatedly genocidal; the men and older women were exterminated, but the young women and children were saved to be used as slaves.

Wars between the Indians and the rubber tappers lasted well into the 1960s. Arito remembers visiting his grandfather Raimundo at the Seringal Iraçema in the early 1950s, when he was about six years old. One night a warring party of Amuacas, blustering and shouting, entered the plantation through a copse of bananas and laid siege to the barracão and the main house for a day and a night. Raimundo and his capangas boarded up the doors and windows and locked the frightened boy in one of the bedrooms. Arito spent the night wide-eyed—listening to the voices and the thud of arrows on the wooden walls. Eventually, Raimundo dispatched one of the capangas to canoe downstream to solicit help. When he returned at dawn with a band of Kampa warriors, the Amuacas fled.

It was a perfect mix for a boom: a cheap, desperate labor force and an endemic product that no one else in the world had. The rubber traders in Manaus and Belém became some of the richest people on Earth. They lived in rambling urban estates behind walls of European tile and sent their laundry to be washed in Paris and London. Both cities equipped themselves with streetlights and electric rail lines. Both imported wrought-iron municipal markets designed by Pierre Eiffel. In 1902 Manaus's Teatro Amazonas opened to the nouveau aristocracy. It had ten-meter-high doors, an air-conditioned theater, and a second-story ballroom ornamented with Italian tapestries and trimmed in Austrian marble. For a few years its linen curtain, twenty meters high, depicting Iara, the aboriginal Mother of the Waters, rose above the most celebrated tenors and divas of Europe (but not, as is widely rumored, Enrico Caruso). Several singers died there of dysentery or fever.

Since the European arrival in South America, western Brazilian Amazonia (including what is now the state of Acre) had been an uncharted hinterland. The border between Spanish and Brazilian Amazonia remained undefined. But when these territories suddenly became valuable real estate in the 1860s, Brazil, Peru, and Bolivia all staked claims. The three nations sent scouting parties to map the valleys of the Ucayalí, Juruá, and Purus. The Serra Divisor seemed to be the natural dividing line. On its northeastern side, the valleys of the upper Rios Juruá and Purus were inhabited by people who spoke a little Portuguese. On the southwestern side of the Serra, the valley of the Río Ucayalí was inhabited by people who spoke a little Spanish. Regardless, in 1867 the government of the Brazilian emperor, Pedro II, relinquished the territory of Acre to Bolivia as a condition of the Treaty of Ayacucho. The Brazilians, it seems, had no comprehension of the magnitude of their concession. The borders of the new Bolivian territory, indifferent to any rational assemblage of tribes, rivers, range, and commerce, spread across the homelands of at least thirty-four Native American tribes (none, of course, were aware of the pen strokes of the mapmakers). The western and southern borders followed the unknown meanders of unexplored rivers; the northern border ran ruler-straight southeast from the Rio Javarí across the paths of twenty-three named tributaries of the Rios Juruá and Purus, as well as several mapped but unnamed streams and *igarapés*. All these tributaries flow northeast, away from Bolivia and toward the Brazilian Amazon, and naturally that was the direction of commerce.

The Portuguese-speaking Brazilians who inhabited the territory had suddenly become squatters in another country, but nothing much really changed. By 1877, 500,000 kilos of crude rubber (about 20 percent of the world's supply) were exported annually down the rivers of Acre to Manaus and Belém. The Bolivian government in La Paz made desultory attempts to manage the territory, establishing a consulate and a customs post in eastern Acre at Puerto Alfonso, on the Rio Acre, to tax these exports. But the consul was practically powerless; the northward flow of the river made enforcement impractical. The law of the land was by no means Bolivian and by no means civil.

In 1899 the Bolivians swallowed their national pride and solicited foreign intervention to reinforce their claim to Acre's riches. A consortium of La Paz businessmen and military brass, under the leadership of General Avelino Aramayo, drafted plans for the Bolivian Syndicate, a colonial enterprise modeled after the infamous Belgian Congo Company. The Syndicate invited American capitalists (including J. P. Morgan and John Jacob Astor) to join. The American government agreed to strong-arm Brazil to allow free passage of Syndicate vessels through its Amazon lands. If necessary, the Americans were prepared to bankroll Bolivia in a war against Brazil. In exchange, they would get 50 percent of the revenue from Acre's rubber production and the right to establish an American colony in the territory.

It wasn't long before news of the Syndicate—still in the planning stages—was leaked to the Brazilians. The hero of the hour was one Luis Galvez, a young newspaper reporter on the staff of Belém's *Província do Pará*. In an act of astonishing naiveté, Chapman Todd, the American navy captain charged with delivering the charter for the Syndicate to the Bolivian Assembly, hired Galvez to translate the document. Instead, Galvez published it in the *Província*. Word of the Syndicate spread westward as fast as the next steamship. Of course, the rubber barons of Belém, Manaus, and Acre had no intention of allowing the Bolivians—much less the Americans—to export rubber from what they considered their territory. In the predawn hours of May 3, 1899, José Carvalho, a local rubber plantation owner, organized a band of armed rabble—the so-called revolutionary junta—to attack the post at Puerto Alfonso and expel the Bolivian consul. It wasn't much of a putsch: the consul and his cronies were roused from bed, escorted to their canoe, and ordered to paddle home . . . upstream.

Within weeks, Galvez himself arrived in Acre and was elected president by the junta. Educated, idealistic, articulate, and steeped in the humanist traditions of democratic Europe, Galvez was hardly one of the boys. Nor was he a Brazilian patriot. No one in the territory had forgotten that Pedro II had abandoned Acre to the Bolivians. Galvez and the junta envisioned an independent, Portuguese-speaking Republic of Acre, which would grow rich and powerful on revenues from its rubber. It would be, he declared, an *expedição do poetas*—an expedition of poets—a utopian nation "which shall have as its foundation liberty and justice."

On July 14, 1899 (a date chosen to coincide with the anniversary of the storming of the Bastille), President Galvez stood on the balcony of the barracão São Jerônimo, a few kilometers south of Puerto (now Porto) Alfonso, and addressed the masses standing in the mud below. He declared:

> It is just that free citizens shall not accept the stigma of pariahs inflicted on them by their Country—nor shall they be able to continue being slaves of another nation: of Bolivia. Citizens . . . We will not continue

to suffer the humiliations imposed on us by a foreign nation . . . We will constitute the independent state of Acre, valiant, strong, and dignified by the patriotism of her sons, powerful because of her inexhaustible riches, which audacious foreigners wish to usurp from us.

Galvez announced that he had written letters to the heads of state in the Americas and Europe declaring the birth of the frontier nation and soliciting their diplomatic recognition. The speech was followed by a grand banquet, for which we still have the receipt.

Lunch for the Revolutionary Junta	860,000 *reis*
American beer, 1 case	180,000
Champagne Veuve Clicquot, 1 case	480,000
Cigars Doneman, 200	200,000
For the public:	
beans, rice, dried meat, bananas, and guavas	
for 107 persons	535,000
Wine Collares	600,000
4 large bottles of cachaça	480,000
Cigars Villar, 500	200,000
Beer Guinness, 3 cases	600,000
Ammunition for rifles, 500	400,000
Lunch for the Members of the Junta still in transit	475,000
Dinner, over 3 days for 11 persons, including	
wine, cigars, and liquor	900,000
Service charges and lost items	500,000
Provision of drinks to the diverse cities that wrote to	
compliment the President of Acre:	
Vermouth, 14 bottles	440,000
Port wine, ditto	260,000
Ginger ale, 6 bottles	48,000
Cachaça, 4 large bottles	440,000
Quinine water, 19 bottles	190,000
Elliptical pills, 50 cases	250,000

After the party was over, nothing changed. The expedition of poets was nothing more than the usual thugocracy of patrões and their enforcers. The ministers of justice, of the exterior, state, war, navy, and police were all wealthy landowners. Native American and Cearense rubber-tappers, marooned in the vast wilderness without any means to escape, remained in perpetual debt.

The independence of Acre lasted only a few months. The rubber barons in Manaus and Belém could not permit the state's economic secession from Brazil any more than they could permit the Americans to seize it. Galvez was deposed in a bloodless coup and exiled to the northeast, his administration replaced by a junta and then a counterjunta. Finally, on February 25, 1904, Brazil formally annexed Acre (for which it paid Bolivia a paltry 2 million in American dollars). The second Brazilian territory of Acre was delineated according to the same irrational borders set forth in the Treaty of Ayacucho.

While Galvez was posturing on the balcony at Seringal São Jerônimo, a separate and largely unrelated struggle for control was being waged by Brazilians 700 kilometers away in far western Acre. There the occupying nation was not Bolivia but Peru. Although Brazilian seringais, including the outpost empires of José and Luiza, were scattered along the entire course of the Juruá, the European settlements on the Moa (which did not begin until about 1895), were mostly Spanish-speaking. Moreover, the Rio Amônea (a tributary of the upper Juruá) was under the martial and fiscal control of the Peruvians, who set up a customhouse at the village of Nuevo Iquitos, on a peninsula of terra firme at the confluence of the two rivers. The bottom line: much of the rubber from the upper Juruá and

Moa was either being taxed by the Peruvians or, worse in the eyes of the Brazilians, being exported upstream into Peru.

The issue was resolved diplomatically on July 12, 1904, when the Peruvians, at a treaty conference in Rio de Janeiro, relinquished to Brazil Nuevo Iquitos and the Rio Amônea as far as the waterfall on the Rio Breu. Brazil had dispatched a battalion of 225 infantrymen to the upper Juruá, encamping them near the mouth of the Rio Moa. (This battalion was soon to become the nucleus of Cruzeiro do Sul.) Upon receiving the news of the treaty, Colonel Taumaturgo de Azevedo, the battalion commander, dispatched an emissary, demanding that the Peruvians withdraw. But Major Ramirez Hurtado, the Peruvian commander of Nuevo Iquitos, had not yet received word of the treaty. The reason was geographical, not political. The Brazilian lines of communication, via the Atlantic and up the Rios Amazonas, Solimões, and Juruá, were far more efficient than the Peruvian lines, via Cape Horn to Callão, over the escarpment of the Andes, up the Río Ucayalí, and by foot past the seasonal rapids of the Serra Divisor. Major Hurtado continued to tax any Brazilian rubber that came down the Rio Amônea, vowing to do so until he received orders to the contrary.

A decisive battle for western Acre was now inevitable. Just before the onset of the rains in 1904, the Brazilian forces, under Capitão Francisco de Ávila e Silva, entrenched themselves in the várzea forest on the left bank of the Amônea opposite Nuevo Iquitos. The force consisted of no more than about fifty infantrymen and forty local rubber bosses, including Dona Luiza, armed with her Winchester .38, fighting at her husband's side. Alas, this revolution was not as benign as Galvez's. On November 4 the two sides exchanged gunfire for eight and a half hours; the next day, for five more hours. One Brazilian was killed and five wounded. The Peruvian losses were not recorded. Soon afterward, about fifty Spanish-speaking settlers were massacred near the mouth of the Rio Moa. Peru's brief but lucrative adventure in Brazilian Acre was over. The Peruvians retreated past the falls of the Rio Breu and, on the Rio Moa, beyond the Serra Divisor.

Cruzeiro do Sul was founded that same year on the terra firme right bank of the Juruá, at the site of the Seringal Estiraõ dos Náuas. It had a barracão, a farinha-processing house, several squatters' houses, and general quarters for the battalions of the operational forces of the Department of the Upper Juruá. Within a few years, thanks to the burgeoning rubber economy, the city became an island of modernity in the hinterland. A post office was opened in 1905, a radio telegraph office in 1912. By 1910 Cruzeiro's population had grown to nearly 2,000, and by 1915 to 3,800; that year there were twelve schools.

Yet by 1904 the whole Amazonian rubber enterprise was already doomed. This change in fortune was directly attributable to an English trader named Henry Wickham, who had been living on the Rio Tapajós, in eastern Amazonia, in a colony of Confederate American exiles from the post–Civil War South, where the exiles were hoping to reestablish their plantation lifestyle—including slavery. The British Empire, at its peak, was acquiring more than terrain, cheap labor, and minerals; it was acquiring the biological treasures of conquest that yielded truly enduring riches. Wickham, seeing the awesome wealth that rubber had brought to Brazil, recognized that England, with colonies scattered throughout the tropics, could dominate world rubber production—if only it had the trees. In 1876 Wickham managed to gather about 70,000 seeds and seedlings of Amazon rubber trees and ship them out of Santarém to the Royal Botanic Garden in Kew, England.

Although a recent in-house publication from Kew declares that "the true story of this enterprise has been widely distorted and popular accounts indicating that Wickham dishonestly smuggled out the *Hevea* seeds are unfair to both the collector and Kew," Wickham himself admitted to bribing the Brazilian customs authorities by "cajoling a friend at court." He wrote, "It was perfectly certain in my mind, that if the authorities guessed the purpose of what I had on board we should be detained."

So Wickham absconded with Brazil's biological patrimony, an action that every red-blooded Brazilian considers an act of international piracy, though Queen Victoria regarded it as a patriotic act worthy of a knighthood for Sir Henry. Only a few of the trees survived the journey to England, but their descendants (some of which are still alive today in the Palm House at Kew) were distributed to English colonial plantations throughout tropical Africa and Asia. It was far more efficient to tap rubber in plantation monocultures than in the wild. By 1913 30 percent of the world's rubber came from British Malaya and Ceylon; by 1920 it was 60 percent. Britain had broken the Amazonian monopoly. During that year the Teatro Amazonas was closed. Manaus once again became a backwater town, slipping into an economic depression that lasted until 1965. Today, as it has been for nearly a century, Brazil is a net importer of rubber.

4

RIVER OF HUNGER

Stars, stars!
And all eyes else dead coals.

— William Shakespeare

WE LEAVE Seringal Aurora at dawn. Now the families no longer wave or smile when we slide by in our canoe. This section of the river is a place of hunger, and they have nothing to trade. One family sprints along a mud beach to flag us down. It is the homestead of Maciel, a refugee from the Transamazonica who is now a squatter on the Moa, tapping rubber at Aurora, hunting and gathering what he can from the forest. Maciel hands me a three-month-old baby, who squeals as she relinquishes the narrow breast of her toothless teenage mother. The baby is his granddaughter. Her head covered in a woolen cap, wearing a necklace with the red cross of Hermão José, she is burning hot.

"She has *doença infantil,* a fever," Maciel explains, begging us for antibiotics. "Is it measles perhaps?"

But to me the symptoms seem much graver: this baby is having her first bout of malaria.

"Since we moved here five years ago," Maciel continues, "all our babies have died of fever before they were a year old." He gesticulates to six small wooden crosses in a clearing behind the kitchen. "Now it is starting again."

Maciel's wife, Jasmina, is boiling a decoction of *tatajuba* flowers in a pan over a wood fire. All her children and grandchildren are wracked with recurring fever, and she hopes that the flowers will cure them, especially the baby. Tarzan scowls at her incompetence.

"Tatajuba makes a good fruit," he explains, "but I've never heard of its flowers being used to treat fever."

Instead, we give the baby an antimalarial, a potentially dangerous drug even in robust people. The drug could kill the baby, but without it the malaria most certainly will. I grind a single white tablet into appropriate dosages for a three-kilo infant. Lacking a scale, I must calculate the dosage by eye, separating the fractions on a page from my journal. I leave Jasmina with a full course of the medicine, enough for three days.

Maciel's family has tamed two sun bitterns and an agouti. "We will eat them when they grow up," he says matter-of-factly, "but now the kids play with them." His yard is covered with hundreds of scarlet macaw feathers, as if a stained-glass window had broken and fallen from the sky.

"Where do all these *arara* feathers come from?" I ask.

"We ate them," Maciel replies. "I used to be able to shoot eight or nine at once, all feeding on a fruiting açaí, with a single blast of the shotgun." He pauses. "But now there are no more."

The newly arrived colonists are "*sem cultura,*" whispers Tarzan. "They are city people. Ignorant. They use the wrong fishhooks. They are always running out of ammunition and beg us for more. They are surrounded by food but will trade their rubber for a case of tinned corn beef. Everywhere they settle, the forest declines. First the monkeys disappear, then tinamous and even the araras."

Maciel tells us that he wants to move his family to Cruzeiro do Sul, where there is a clinic, a school, and a store with fair prices. His oldest son, now eighteen, doesn't want to spend his life on the river. But Arito cautions him about the city's hostility.

"What kind of work could you find there?" he asks Maciel. "Do you have relatives in Cruzeiro?"

"None."

"Where would you live, then?"

"I don't know. But could it be worse than here?" Maciel complains.

"Yes," Jasmina reminds him. "We came from Belo Horizonte, from the favela. How can we return to a city?" She sighs. "Is there any place without bosses?"

We can't do much more for these people. The baby will probably survive this bout of malaria, but even under the best of circumstances she will become infected again. Tarzan buys seven *jabutís*—red-footed tortoises—from Maciel and stores them in the prow of *Fe em Deus*. We don't really need them or want them, but the purchase gives Maciel's family a few farthings.

At the next homestead we find more misery. A man is lying on a bed of fresh-cut leaves in a regatão's open canoe. His wife seems dazed; a toddler hugs her skirts, dumbly watching his father. She tells us that three days ago at dawn, while cutting rubber, he was bitten on the calf by a snake. He didn't see the snake in the estrada, but it was probably a *surucucú*—a fer-de-lance, which has the color and texture of the leaf litter. By the time the man staggered back into camp, his leg was discolored by hemorrhage and was swelling.

His wife didn't have enough strength to haul him to the canoe, so she waited three days until the regatão appeared.

Now his leg is twice as large as usual, and the skin is splitting open. She asks us for help, but we have nothing to offer. The venom has deprived his blood of its ability to clot, and any injection or anti-venin at this stage could cause him to bleed to death. She squats on a board by the canoe, rocking on her haunches, sobbing, wiping tears on her dress. Everyone is waiting for some sacks of farinha to be loaded onto the canoe. The farinha is the reason for the journey to Cruzeiro, not her husband. I look into the eyes of the dying man, but he is addled now, incoherent, awed by the impending darkness. No sentient being is looking back at me.

"It is as the rasga mortalha predicted," says Tarzan. He is relieved. "At least it wasn't one of us."

Tarzan has known the Rio Moa most of his life. His memories of this dynamic landscape are older than its present meanders and sandbars; he remembers places that no longer exist and people who are long dead. He was born in 1944 in Tarauacá, only 180 kilometers from Cruzeiro do Sul as the crow flies, but in those days it was a river journey of ten days down the Rio Tarauacá to its confluence with the Juruá, then upstream for ten more days. Tarzan's father, Paulo Fernandes Derzi, came to Brazil from Syria in 1941, recruited as a soldier in the *Exército da Borracha*—the army of rubber. After the Japanese captured Indochina, Malaya, and Indonesia, the Allies lost their principal source of rubber. An army marches and rolls on rubber, so the Allies looked to Brazil's native rubber trees. The *Exército* recruited tens of thousands of desperate, poor men, mostly from Brazil's northeast but also some foreigners, to tap the wild rubber trees of Amazonia. They soon became hopelessly indentured, their consumed by disease, accidents, and fights, with no escape the seringais.

Tarzan does not know whether his father was born in the desert or the city nor whether he was Muslim or Catholic. Regardless, he was not suited to the conditions in Amazonia and died during an epidemic of malaria when Tarzan was three. His mother, Raimunda, quickly remarried, but the stepfather beat the little boy. Nobody objected when Tarzan ran away from home at the age of six and became a *peixote,* sleeping on the streets and begging for food. After a while the child was beguiled by Lindava, the wife of an air force sergeant, who smuggled him among boxes of freight on a plane to Cruzeiro do Sul. Lindava put him to work selling cakes and soups in her lunch stand. It was exploitative, but at least Tarzan slept under a roof and wasn't beaten. But soon Lindava, too, abandoned the boy; she bartered him to Maria Rosa, the gaudy owner of the town's most prosperous inn, the Hotel dos Viajantes. Tarzan slept on the lobby floor and earned his keep by heating bathwater over a wooden stove and hauling it in twin five-gallon tins yoked to his shoulders to the patrons in the rooms upstairs. The work started at three in the morning and continued until midday. By the time Tarzan went to sleep in the late afternoon, he had carried as many as 170 or 180 tins of water.

When he was thirteen Tarzan fled Maria Rosa and hitched a canoe ride with a party of hunters, three days up the Rio Moa, then five more days up to the far reaches of the Rio Azul, to the Seringal Xambira. A city boy, he had no experience in the forest, but hauling water had made his frame lean and hard, and he had no expectations of kindness. He got none. Kelly, the patrão of Xambira, gave Tarzan an aviação consisting of a rifle, some ammunition, a rubber cutter's knife, eighty tin cups, a sack of farinha, some coffee and sugar, and a license to tap eighty rubber trees. In exchange, Tarzan was obliged to give 40 percent of his bolas to Kelly and to barter the remainder to the company store to repay the loan.

Tarzan built a hut of *paxiúba* palm thatch on the terra firme at a bend of the river, and from there threaded two estradas, each about ten kilometers long, through the closed forest, over terra firme and várzea. He worked the estradas on alternate days for four months straight, letting the trees recover the rest of the year. It was relentlessly tedious work. Before dawn, when the sap of the rubber tree runs fastest, he set out barefoot on the gloomy estrada, his way lit by a sputtering oil lamp, known as a *piraqueira,* strapped to his forehead.

"Sometimes," Tarzan told me, "in the lurching light cast by the piraqueira, the shadows seemed to jump out at me. I was sure I was being followed."

After setting the cups, Tarzan had to walk the estrada once more to collect the latex, returning to the camp at three or four in the afternoon. But his work wasn't over: he still had to mold the latex into a bola over the acidic, coagulating smoke of a fire of mata mata wood for five or six hours. He would cook his meal over the same fire before tumbling into his hammock.

One night, while retracing his steps to pick up a machete that he had dropped, Tarzan noticed that a jaguar was following him, placing its pug marks exactly inside Tarzan's own footsteps in the wet mud. The big cat must have been only a few meters away. The boy hid between the buttresses of a tree and waited, rifle cocked, for the dawn.

"On the next night," he said, "it took all my courage to return to that forest. But I learned to swallow my fear. And I've never been afraid of anything—or anyone—since."

During the rainy season, when parts of his estrada were under water and the game animals take refuge in the pockets of dry upland forest known as *terras ilhadas,* Tarzan hunted. But it went poorly at first; during his first three months, he caught only one small ocelot. His supplies were exhausted. For a week he subsisted on nothing but the fruits of the *pataúá* palm and some dried monkey meat.

"I became so weak that my legs were wobbly."

And it was an unhealthy forest. Twice the crusty red lesions of cutaneous leishmaniasis—a potentially lethal protozoal infection that results from the bites of the minute *tatuqueira* fly—erupted on his legs.

Tarzan endured and slowly learned on the job. He began to acquire jeito. He cured his leishmaniasis with poultices of ground lead shot, *gomeleira* sap, and *copaíba* oil. He crafted an ocelot trap from the round prop roots of paxiúba barriguda trees, inside which he suspended a log of *canelão de velho,* as heavy as an iron ingot. He baited the trap with dried smoked meat of a howler monkey, white-tailed deer, or brocket deer. When the cat tugged at the piece of meat, the log was released and smashed its head. On one day alone, Tarzan caught five ocelots, and within a year he had cached three hundred cat skins, two hundred redfoot tortoises, and about a ton of dried meat.

One morning Tarzan found two crossed macaw plumes in the middle of his estrada, a warning that he was trespassing. He never learned who placed them there.

"I figured that they were Amuacas," he said. That was the tribe, legendary for its resistance to rubber tappers, that had laid seige to Arito and his family at the Seringal Iraçema twenty years before.

Tarzan decided it was time to cash in his chips, and the next day he left the forest for Xambira. The boy took his booty—bolas, skins, tortoises, and smoked deer and monkey meat—to the barracão for exchange. They were worth more than enough to pay his rent

and repay his aviação. Flush with prosperity and delirious wit
camaraderie of the seringueiros, Tarzan bartered a bola for si:
tles of cachaça and, in the course of the afternoon and evening,
shared them with the others.

"I became *doido*," he said, "but was I ever proud."

He passed out, and when he awoke, he was lying, encrusted in
dried vomit, on planks above the pig sty. His rubber, skins, tortoises, and dried meat had all been stolen, and the barracão was de-
serted. Tarzan waited a night and a day before hitching a ride downstream on a passing canoe.

By the age of fifteen Tarzan had made a fortune and lost it on
the Rio Azul.

Back in Cruzeiro do Sul, Tarzan took a job in the brick kiln, stamping the red clay with his bare feet. But the humdrum job didn't suit
him. He had jeito now. Soon, with his savings from the brick factory, he salvaged a small canoe, bought some paddles, and went
into business with Gilberto Japiní, a seringueiro from Paraguacú,
three days downstream on the Rio Juruá, who had also managed to
escape his debt. The men worked as fisherman on the upper Juruá
and its tributaries, and as salvage divers, disentangling fish nets in
exchange for part of the catch.

"It was a dangerous job," Tarzan once told me, "especially in the
rainy season, when the fast current can push you into the net or a
logjam and drown you." One of Tarzan's specialties was removing
electric eels and sting rays from the nets. "The poraqués were easy.
Once we got to them, they'd lost all their electrical charge. But the
raias were dangerous. We'd find five or six of them at a time. You
hold them by their mouths—which suck but can't bite—and cut
their tails off so they can't slap you with their barbs."

Tarzan and Gilberto were too poor to buy a net. In fact, many of the fishermen didn't use nets. It was far cheaper to hang a lantern over the water at night and lure the surface-feeding species, particularly the *sardinha* and *peixe cachorro,* to jump right into the boat. And during the months of the piracema, when there were more fish than the canoes could hold, Tarzan and Gilberto found a much more profitable business: selling salt to preserve the fish. It was far more valuable than either gunpowder or quinine.

After a few years Tarzan had acquired sufficient wealth to build a two-room house on an unclaimed lot in the terra firme outside of town. He married Gilberto's sister, Geni; they were both seventeen. Tarzan took up a more domestic life, selling fish, salt, chickens, and bananas on the street in front of the market—which was against the law in the 1960s, when the military government awarded commercial licenses in exchange for bribes. The police arrested him, and within hours a magistrate sentenced him to two months in jail. After his release, Tarzan shifted his business to his house. But soon his entire squatters' neighborhood was condemned; it was to be torn down to build a new radio tower. Tarzan refused to move. One dawn a sergeant and three military police overpowered him, bound him, and dragged him away. The house was razed by noon, and Tarzan, having no formal title to the property, was never compensated.

It was a tough time for Tarzan. He repeatedly got drunk, started fights, went to prison five or six more times—he doesn't remember how many. On one occasion the military police beat him up and broke three of his ribs. But Geni pulled him through. She had adopted a niece, Katia, and the pressures of paternal responsibility brought Tarzan around. The couple staked a claim in the Bairro Flutuante, on riverfront that nobody wanted, which meant it was safe from the bureaucrats and the tax men. They built their house during the dry season atop two assacú logs that Tarzan had fished

from the river. When the flood came, the house simply rose
occasion. Assacú wood lasts fifteen or twenty years. But by th
the logs under Tarzan's house had started to rot, the Bairro Flutuante had grown prosperous, with wooden sidewalks built on stilts,
electricity, and, with Brazil's return to democracy, legitimacy. One
winter during the flood season, Tarzan simply roped his house into
the scaffolding of sidewalks, planted some stilts, and pushed away
the logs. He had become a member of the landed gentry.

At dusk we reach a wide bay fringed by capim. It is the mouth of
the Rio Azul, the tawny tributary of the Moa that will take us to our
research site. The Azul is too narrow and full of snags to be safely
navigated at night, so we set our anchor. Rather than build a camp
in the darkness, we will sleep on the *Fe em Deus,* tying her to a log
in the middle of the river and stringing our hammocks from her
rafters. Geni fires up her red stove and cooks fried plantains, a slab
of bacon, some rice, and beans, ordering us to stand clear and leave
her alone.

Taking the hint, Arito and I each take a dugout canoe, paddle
hard upstream for an hour, then drift silently, allowing the current
to carry us back in the swirling reflected starlight. A well-crafted
dugout canoe is otter-swift and so exquisitely balanced that it becomes an extension of your body. In the moonless night, undis-
tracted by the day's bright visual clutter, I become the river. I can
feel the water through the body of the canoe, the eddies, the up-
welling snags, even the wakes of fishes. And a well-wrought oar—
lightweight, planed from a single strong buttress, so perfectly tear-
shaped that every stroke is effortless—merges with the river as if it
were itself made of water.

As we drift, a million ears are listening, a million eyes are open tonight, scrounging every last photon of starlight for a hint of food . . . or danger. A splash, the bark—*unk! unk!*—of a spiny tree rat, the three-note melody of an owl. Hoatzins are hissing and huffing in a copse of mungubas. A dozen caimans croak—*ankh! ankh!*—on either side of a meander. Afterward, a long silence: the patience of night-things.

Now we drift among a pod of *bôtos vermelhos*—pink dolphins. They are feeding on bottom fish, fluking up eddies that modestly alter the trajectory of our canoes. The pursuit of dolphins is like the pursuit of beauty: you never quite attain it. They are as fey as rain. You have no choice but to wait for them to come to you on their own terms. But dolphins are highly intelligent, and you can exploit their curiosity. Arito whistles—two or three descending notes, then a long zitherlike ascending one—to attract their attention. No reaction. The whistle is an old wives' tale, and I'm surprised at Arito's gullibility. But I try another trick: slicing the water with the sharp edge of my machete; the rapid shearing sound seems to mimic their whistles. The bôtos drift with us now, their curiosity piqued. I still can't see them, but I can hear them and I can smell their halitosis. A mother is diving with her calf, breathing in synchrony with it. Each of the others has its own characteristic blow: raspy like a wheezing old man; wet; brief; tight-sphinctered; bubbly; belchlike.

If only I had the gift of night vision, a cat's eyes to peer along the river's edge and into the dark heart of the forest. Just about every creature in this forest and river, save for me, has a tapetum, a reflective layer at the back of the eyeball that bounces light a second time through the sensory rods and cones of the retina to amplify it. It is the tapetum that gives nocturnal animals their "night vision," allows them to turn night into day. Silvery retinas have evolved independently in a variety of nocturnal animals, including shrimp, fish, amphibians, reptiles, birds, and mammals. Alas, my ancestors were diurnal, spending their nights huddled in fire-lit caves, and I am not so gifted. But, being a technological beast, I can see these animals as well as they can see me if I hold my flashlight precisely at eye level, at a right angle to my face, so that the angle of incidence, the angle of reflection, and my line of sight are all the same. Now every tapetum in the forest becomes a mirror, every retina a lantern. Each species has a signature reflection, and there is a certain taxonomy of eye shinings. The eyes of spiny tree rats reflect a metallic yellow; they are mouse-alert, always looking around. Any bird, if you are patient, will eventually blink. The eyes of nightjars first reflect blue-green, then, as you approach, red; they hold their heads steady, but you can see the retina bouncing inside the eyeball as the bird surveys its environment. Owls and potoos like to perch on dead snags to survey their territory; they move their heads almost 180 degrees, blink, then survey again. Lizards have eyelids, and they either blink or lick their eyeballs; in either case their reflection momentarily pauses. But a snake's eyes are always open, unblinking, steady.

I sweep my flashlight across the capim, through the opalescent strands of spider silk slung over the blades of grass and lofting in the wind. The beam intercepts a couple of fish-eating bats grabbing long-fingered handfuls of air, enabling them to change course in an instant. A thousand cold blue diamonds: spider eyes, easily confused with drops of water on the river's surface. But if you look closely, you can tell the difference: the spiders lie flat against the capim culms and the bases of inundated trees, waiting for the mayflies to emerge. Below the surface, cold constellations of freshwater shrimp, transparent except for their opalescent eyes, fidget with the vegetation. A disoriented *acaba da noite*—a night wasp—blunders toward the light, travels down the beam as if it were a highway of photons, blunders onto my face, and falls inside my shirt. It gives me a painful but short-lived sting and an itchy welt that will last for days.

The sardinhas are also gorging on the mayflies. True flying fish, they are able to glide above the water on two triangular pectoral fins. But tonight, alarmed by our passage and confused by our lights, they fly into the boat, slapping me like an animate silver hailstorm. Instinct tells them that safety is in the air, but instinct did not prepare them for humans. A needlefish approaches the canoe, its tweezer-face nipping at the insects drawn to our lights. The fish is one-third head; its lobed eyes provide forward-looking stereoscopic vision. Its hindmost vertebrae are crooked so that its caudal fin can pull it in reverse.

Now we drift past a tangled bank of trees toppled into the water. Iguanas roost here, immobile, claws dug into the bark. Diurnal, they lack reflecting eyes and so are invisible, but as we approach, they give themselves away, plopping into the water like ripe fruits. The prey of hawks and eagles, iguanas instinctively hide in the river. Using my paddle, I flip one into Arito's canoe. He lurches, grabs it, almost capsizes, tosses it overboard, and cusses.

Ahead, a hollow snag is releasing a fetid, warm column of bat body odor and the ammonia stench of guano. The column has a different refractive index from that of the night air and is visible in the flashlight beam. This tree, and its bats, belong to a three-meter-long snake known as a *surucucú rana,* which is draped over a branch like a pudgy tendril. Such easy pickings: this snake is fat-bellied, digesting yesterday's bat.

"Cuidado," Arito cautions with mock gravity, as the prow of my canoe bumps the tree. "It has enough venom to drop a horse."

But I know better. Arito is paying me back for tossing him the iguana. The surucucú rana is no more than a prehensile-tailed boa, a constrictor whose thin neck, triangular head, and diamond-patterned back mimic a pit viper well enough to give a hawk second thoughts. And I as well. Who can be sure it's rana? I push off.

All along the shore, the caimans are barking . . . a nasal *unkh! unkh! unkh!* We call back—Arito and I are both good caiman impersonators—and slam the broad faces of our paddles on the water to mimic the way they slap their tails. In an instant the shoreline is bright with a dozen caiman eyes—reddish yellow embers in the reedy sides of the meanders, as if a string of Christmas tree ornaments suddenly clicked on. They are all juveniles, and most are *jacarés tinga,* spectacled caimans, the most abundant species on the Juruá and its tributaries. Seldom exceeding two meters, the jacaré tinga is shy and docile. We also spot a *jacaré tinga cascudo,* a smooth-fronted caiman, with a dull yellow eye, hiding in the roots beneath the undercut, eroding side of the river. At a little over a meter in length, it is the smallest but the most aggressive of the caimans.

"I never lost my respect for them just because they are small," notes Arito.

So far tonight we have seen no sign of the black caiman, *jacaré preta,* the largest species. An adult can grow to eight meters long and a meter wide—bigger than our canoes. It is one of the most formidable crocodilians on Earth, but individuals that size are rare now. The jacaré preta loves the snarled, closed vegetation of the várzea, especially fallen tree trunks, from which it can ambush its prey. Fiercely territorial on the nest, a mother black caiman can swim as fast as a torpedo and swamp a dugout canoe, swiping it with her tail, butting it with her head, or even crushing it in her jaws.

Suddenly the left bank fills with a dozen jacaré preta eyes. They are easily distinguishable from the other species—dark wine ruby red, like cooling cinders—and these particular eyes are all tiny. Hatchlings.

"Uh oh!" says Arito. "Where's mama?"

⚬

The start of the rainy season—just about now—is when the caimans hatch. About two months ago, their mothers built layered mounds of earth and vegetation in which they buried their eggs. The amount of rotting vegetation relative to the volume of mud must be just right, because the heat of decomposition incubates the eggs. If there isn't enough heat, the eggs will develop slowly and drown in the floodwaters before they hatch; if it's too hot, they will cook. The mother monitors the warmth of the mound by resting on top of it, using her sensitive white belly as a thermometer. By sliding her wet body over the mound and by urinating on it, she is able to adjust its moisture and therefore its temperature.

Now the eggs are beginning to hatch. The babies start to bark while still inside the mound—an extraordinary plea for assistance during their short but perilous passage to the river. Cued by their cries, the mother tears the mound apart, lifting the newborns in her jaws and carrying them with great tenderness and delicacy to the river. The hatchlings, only ten centimeters long, eat insects and other small fry. It is dangerous to canoe on the river at this time of year because mother caimans, fiercely protective, guard their babies for several months after they hatch. Regardless, most of the babies will be eaten—by herons, fish, anacondas, other caimans—before they are a year old. Only when the survivors are about thirty centimeters long do they become invulnerable to most of their enemies.

Arito is a good companion in caiman country. He can think like a jacaré, can anticipate their moves. He learned these skills in the

1960s, when he hunted caimans on the Rios Juruá, Japiim, and Moa. He worked in the dry season, when the caimans were concentrated in the belly of the river; during the winter, when the river spilled into the forest, they were too widely dispersed to make hunting them worthwhile.

The hides of the smaller species had limited value. The jacaré tinga is fast-growing but small, with thin skin that often splits when it is cured, and the jacaré tinga cascudo has too many armored dermal plates—there are plates even in its eyelids—to make fine leather. The market paid well only for the jacaré preta. And the pay was extravagant, at least in the context of the local economy. The value of a single two-meter hide, sold in Cruzeiro do Sul or Tabatinga, was 90,000 cruzeiros, the equivalent of a month's labor on a seringal.

Arito, living on a vessel the size of *Fe em Deus,* hired a team of three men. They slept by day and managed to kill eighteen or twenty caimans a night. Arito was the *arpoador,* the harpoonist. The shaft of an *arpão* is tipped with a two-toothed barb tied to a float known as a *cortiça,* made of buoyant balsa wood. Fixed to the end of a lance, the float is designed to separate from it once the caiman is impaled. Like the harpoonist on a whaling ship, the arpoador was the most valued and highly paid team member. Standing in the prow of a dugout while one of his buddies paddled from the stern, he held the lance in one hand and one of two spotlights in the other. A light with strong new batteries was used to shine the caimans' eyes from afar; another light, with worn old batteries, was used to sneak up close. It took consummate skill to stand in the prow of a dugout, throw a harpoon with enough force to impale a caiman, and not capsize the canoe. The maneuver required the arpoador to instantly shift his weight to the foot opposite the throwing arm.

"It is best to spear the jacaré in the neck, just forward of the front arm," Arito explained. "Then she will roll over and over, wrapping the cord around her legs and body. Eventually she will immobilize herself, and be held afloat by the cortiça. All we had to do was shoot her."

Caimans, especially the older ones, are generally shy. "But when jacaré preta builds her nest during the summer, she becomes doida and will attack people without provocation," cautioned Arito. "When you can hear her babies croaking—stay away. Once one of my friends, Leonidas Pessoa, tossed a cast net on top of a baby jacaré preta. The mother came from nowhere and bit off his right arm."

Arito has had two close calls with black caimans. Not far from here in 1969, a female six or seven meters long who was guarding her nest on the shore lunged at his canoe, crushing it with her jaws.

"I kept my balance on the broken prow of the canoe and killed her with a machete chop to the neck."

Another time, at the Seringal Condor on the lower Juruá, Arito cut the neck of a three-meter-long jacaré preta and, thinking it was dead, hauled it into his canoe.

"It wasn't dead. It came to life, biting and rolling. Everybody jumped into the water."

Hunting caimans was outlawed in 1972, but by then the big ones were becoming scarce, and although the prices were elevated by their rarity, it was no longer worth mounting an expedition to hunt them. Their numbers have not recovered. In spite of protection, the jacaré preta is still rare on the upper Juruá.

Tonight we have a problem: the legion of jacaré preta eyes in the meander ahead. The mother must be nearby. I am chagrined. It's stupid to be in this situation, a cliché, really: a mama caiman between us and our rice and beans. We didn't see all these babies during our trip upriver because we were on the opposite shore, too far away to scan their little eyes. But why didn't I see the shine of the mother's eyes as we passed?

"Obviously," says Arito, with a hint of impatience, "her eyes were shut. She must have been asleep on top of her nest. I hope we didn't awaken her when we paddled upstream," he muses, "because now she'll be waiting for us."

We paddle against the current, remaining stationary, while we figure out a strategy. We decide to drift past the caiman in abject silence, not stirring the water or scraping the paddle against the wood or bumping the side of the canoe, or ever so slightly shifting our weight.

"She'll think we are only a log," declares Arito.

Of course she knows better. Our scent may have cued her, or a growl of intestinal gas. A splash—as loud as a falling log—somewhere on the right bank. After a long moment, my canoe rises on her bow wave as she bolts underneath it. And then I spot her, two meters astern where she has breached and turned to face us. Her head is a meter and a half long, her eyes, as bright as burning coals, are as big as tennis balls. Their reflections are red bloodstains in the water. She arches her tail, serrated with scutes as big as my hand, and slaps the river, then bellows. The water resonates, dances as if pelted with rain.

"Paddle!" rasps Arito.

But I can't take my eyes off her, this dark Mesozoic nemesis, the color of the star-collied night. She could take us if she wished. With one stroke of her tail, she could be upon us. But she is a mother, and good mothers are conservative—why risk injury to confront an intruder who is already fleeing away? The caiman and I watch each other until we drift around the next bend.

"*Puxa*," sighs Arito. "In the name of God, don't tell Geni."

Fe em Deus is tied to a snag in the dark water, a silhouette in the diffuse starlight that seeps through a newly risen veil of mist, back-lit by the Milky Way: gauzy wads of galaxies and stars and glowing gas, marbled with the dark matter that is starlight's shadow. A silent green meteor crinkles the heavens, an ephemeral trail of green smoke. Now the cat's-claw moon dangles on the limb of a samaúma. We can hear a few *bacurãos*—nightjars—purring. Nothing more.

5

WHAT THE BEES SAID

These are true mornings of creation, original and poetic
days, not mere repetitions of the past. There is no lingering
of yesterday's fogs, only such a mist as may have adorned
the first morning.

—Henry David Thoreau

THE DOG is nudging my hammock at 5:13 A.M., but I linger, steeped
in the purple dawn. There is a stillness, a pause, before the sun ap-
pears. The night animals are hieing back to their burrows and
folded leaves, their bellies full. The night sounds, which seemed so
ardent, so imperative, are silenced, and the day sounds have not yet
begun. Only a few crepuscular *Anopheles* are singing outside my
mosquito net.

At first light I walk to the river's edge and splash my face, holding the cold water against my eyes. A membrane of fog hangs above
the Rio Azul, wedged between layers of cool night air. The shore of
the Juruá has seen a billion dawns, each as original as the Earth's
first dawn. On the river a pod of *bôtos* leave chevron wakes; their
clouded breaths linger long after they've sounded. Three large-
billed terns dip into the river's still surface, nabbing the top min-

nows flushed by the dolphins. The early bees are fussing at the *flores de mulher*, sucking the first nectar and tucking away the morning's crop of pollen into pouches on their legs. In a ucuüba above, a toucan is yipping; it sounds like an abandoned puppy, lonely and forlorn. The low-frequency murmur of a curassow, almost too low to be heard, like bones rubbing. The rusty-hinge squeal of a flock of mealy parrots. Somewhere a woodpecker is extracting an ombrophilous beetle larva from the safety of cellulose and lignin. Now the ululating gibber of the family of dusky titi monkeys that live in the várzea forest across the river. After a few minutes, when the sunlight has leaked onto the low mist, the kiskadees cry, *"Bem te vi, bem te vi."* The cough-barks of a spiny rat. And then, from across the river comes the hysterical, raucous call of a flock of *aracuãs*. Caboclo children are taught that they are crying the onomatopoeic *"Todos velhos feo, todos velhos feo, afora meu avo, afora meu avo."*

The quiet of the forest's awakening—the day's big yawn—doesn't last long before Geni's radio intrudes. The programs are daydreams: the reverba voices of mock-enthusiastic announcers selling city products totally irrelevant here—Nescau, dishtowels, Vienna sausages—playing polkas squeezed from an accordion, announcing obituaries with droning genealogies of survivors and personal messages to wanderers and rubber tappers. "Nielson Sousa, your mother is comatose; your sister Renata requests that you return home immediately." This is the medium of modern Amazonian culture, the fabric of a wilderness frontier.

The dog wandered into our camp a few days ago, starving and infested with botfly maggots. No doubt he became lost while hunting or was abandoned on some lonely homestead when things got

tough. We've named him Tatuzinho, and he curls up next
warm stove. Always glad to see us, he's a polite little camp
tle and deferential.

As the sun rises, the heat volatilizes the aromas and stenches of
the forest. The warmth descends slowly from above; the light, too,
percolates through the understory as the sun climbs higher. For a
few minutes the layers of atmosphere are shuffled, and the night
and day aromas mix: the urine spray of an ocelot, a yeasty egg-laden
frog spume, musk rubbed from the anal glands of a jaguarundi,
the tang of a fruit, the fleetingly fetid stench of a decomposing cat-
bird. I am lassoed by a wandering tendril of orchid aroma reaching
through the forest, its flower unseen somewhere in the canopy
overhead—just a few well-placed molecules but sufficient for the
economy of day-flying bees. The forest is in fact a cornucopia of
smells, but I, lacking the olfaction of a jaguarundi or a bee, am in-
sensate. We primates conceptualize with our eyes, and although this
skill served our early ancestors, who had to spy big cats in the tall
grass, it blinds us to the true signals of this forest. Unable to decipher this chemical language, I am only fleetingly aware of a syllable
here and there.

Now the aroma of Geni's breakfast wafts past: fried piranha
preta, a bland mixture of farinha and water called *shibé,* and coffee
so sweet and strong (the grounds are boiled in brown sugar and water) it makes your teeth ache. The dull thuds of splintering wet bone.
Tarzan is slaughtering one of Maciel's jabutís, splitting it, still alive,
with an ax. It would be more humane to decapitate it first, but that
is impossible because, with every blow of the ax, the tortoise with-
draws its head farther. Her translucent oviduct is full of orange-
yellow eggs, each one larger than the preceding. The tortoise's intes-
tines twist in the unnatural light, tugged by violated mesentery. Now

her dismembered head, neck, and right foreleg, still attached to one another, fall to the ground, the eyes wildly uncomprehending.

⬠

We have been camping at a place named Valparaiso, on the edge of the Rio Azul, for a week. It's a lovely spot, high and dry from the flood but close to the nourishing várzea forest. Few people live at Valparaiso these days, but there is abundant evidence that it was once inhabited—a jambu tree, a papaya tree, a feral banana field, some scattered manioc plants left by seringueiros who abandoned this spot years ago. Before the seringueiros, Native Americans must have lived here, but their only traces are a few shards of pottery, some stone ax heads, and *terra roxa*—the signature russet earth left by their kitchen scraps and latrines.

Along the Rios Juruá, Moa, and Azul are thousands of ruined homesteads like Valparaiso. The people who lived in them left no enduring trace. What stories do these forgotten places hold? Who lived and died here? Who cut rubber in this forest, nurtured children, watched them die, farmed this place, turned the soil red? Who cut the estradas through várzea and terra firme, learned every nook and nuance of these trees?

Years after my first visit to the Rio Azul, quite by coincidence, I learned the answer to those questions for this place. In 1984 I was spending a month at Santa Luzia, a pioneer settlement on the Trans-amazonica fifty-nine kilometers east of Cruzeiro do Sul, conducting an inventory of the trees and collecting plants. In those days the Transamazonica was sequined with brand-new villages—known officially as *agrovilas*—every fifty to a hundred kilometers. Santa Luzia was one of them. They were the hope of the new frontier, nuclei of Western modernity from which civilization would inevitably spread throughout the region. Each agrovila was, in a sense, a kernel of seed corn, and each represented a huge investment by the government in Brasília, which paid for a generator, a radio transmitter, a well, a school, a warehouse for the anticipated agricultural bounty, and, of course, white-collar administrators to oversee it all. The agrovilas were conceived as convenient places to settle the burgeoning populations of refugees from the drought-stricken farmlands of northeastern Brazil and from the urban *favelas* of the southeast. The government seduced the destitute from those areas, offering hope, a new life along the marvelous highway to the New World, and, above all, land.

And Santa Luzia was the promised land in 1984. Perhaps a hundred families lived there. About half were refugees displaced by the flooded reservoir of the Itaipú Dam, 2,000 kilometers to the southeast on the border with Paraguay. Each family had been given title to a homestead of five hectares of terra firme forest, an ax, a machete, a few kilos of nails, some tools, and sufficient boards to build a house. Big-time investors had moved in, too, clearing vast expanses for cattle ranches along the highway, pushing the wall of trees over the horizon. The settlers of Santa Luzia considered themselves pioneers and were buoyed with pride. The schoolhouse, made of neat boards with a tin roof and wide, breezy windows, was filled with children dressed in starched white shirts. Using government subsidies and loans, the colonists had installed a generator, dug a well that all could share, built a clinic, hired a nurse. There were stacks of fresh-hewn lumber, still aromatic with sap. In the administrative building clerks typed forms, stamped receipts, authorized payments. Heaps of letters awaited the signature of the director. In other words the office had *fiscalização,* the overwhelming, yet comforting Brazilian bureaucracy that gave day-to-day activities an air of authority and permanence. The building was air-conditioned, so people hung out there whenever they could just for the novel sensation of feeling the lovely cold and watching the marvel of big-city paperwork.

The settlers, mostly from other parts of Brazil, were naive and lacked jeito, yet somehow were expected to hew a living from their clearings in the forest. But the government, instead of encouraging the colonists to plant subsistence crops adapted to the terra firme soils, such as manioc, beans, papaya, and bananas, ordered them to grow a cash crop: onions. They were given no choice. All of Brazil was monitoring the agrovilas. Santa Luzia was wholly dependent on government funding, and the director knew that the government's investment could not be repaid with subsistence farming. Besides, he anticipated making a killing with the onion crop in Cruzeiro do Sul; onions had traditionally been imported from the south and therefore were seasonally in short supply. So Santa Luzia's principal enterprise became raising onions in the terra firme soils.

We lived and breathed onions that summer. The director allowed us to string our hammocks in the shed where the onions were stored, and after a few days our clothes, hammocks, and bodies all smelled of the sulfurous bulbs. Nobody could stand to be with us for long. Yet the warehouse was largely empty. A few dozen gunnysacks of onions—hardly worth the price of the director's air conditioner—was all that the settlement produced that year. Most of the onions were withered and streaked with a slimy black fungus. Nobody wanted to buy them.

One evening a settler named Joaquím stumbled into our shed, shaken, with his wife and two children. Afraid to return down the dark trail to his hut in the forest, he begged to stay with us for the night. He was a new recruit from a favela in Minas Gerais, but he had already lost hope. Acre had turned malevolent to him.

"I thought I would be given good land," Joaquím complained,

"land that I would be proud to pass on to my son. But nothing grows in these soils. The onions rot in the very earth that should nourish them."

Earlier that evening Joaquím had gone out to the outhouse behind his garden and had laid down his piraqueira just outside the door. Suddenly a *pico da jaca*—at two meters long, the largest of the pit vipers—struck and knocked over the lantern, snuffing its flame. Joaquím waited in the darkness, listening as the snake retreated into the yard.

"Deus!" Joaquím shuddered. "If I had not set down my piraqueira, it might have bitten my hand."

The next day Joaquím decided to consult a grandmother named Dona Cabocla, widely regarded as the matriarch of Santa Luzia. She had lived in the settlement since 1976, four years after the experiment began. Unlike the other settlers, Dona Cabocla had been born in Acre; she had lived on its rivers and toiled in its forests. In short, she had jeito.

"Dona Cabocla has a way with things," Joaquím informed me. "She will know how to rid my garden of the pico da jaca."

And so at dawn Joaquím and I visited Dona Cabocla. She was a little over five feet tall and as lean as a stick. A red hibiscus was folded into her long gray hair, fastened with a frayed blue toothbrush as a barrette. Her expression was slack, almost sorrowful; she had a furrowed, aquiline nose, downturned lips, and eyes that seemed worn out by what they'd seen. But when she spoke, Dona Cabocla looked at you straight on with a piercing intelligence. She practically sang, gesticulating with long crooked fingers, and would clap her palms and exclaim "Oh! *Meu Deus!*" as if anything worth saying should be spoken voluptuously or not at all. It was clear to me that this woman was a creature of light. Everything she touched was nourished, everything she planted grew: her sunny green *rocinha* (home garden) of manioc, bananas, rice, papaya, and jambu as well as the cats, dogs, parrots, pigs, and chickens, and the wanderers who sought shelter in her homestead. And, of course, her children—all adults by then—and nine grandchildren.

As expected, Dona Cabocla had a solution to Joaquím's snake infestation. She listened to his story attentively and compassionately. Then she rose and pointed a long finger at him, as if scolding one of her children.

"Well, your mistake was to carry the piraqueira in your hand," she explained. "You see, the pico da jaca doesn't see any better at night than we do. But it feels heat. It must have thought that your lantern was the warm body of a rat or an agouti. Now is the beginning of the rainy season, and the snakes are roaming about after sleeping all summer, and they are hungry. Always strap your piraqueira onto your head or hook it to the end of a stick.

"*Não problema,* Joaquím. Here's what you do," she went on. "Arrange some stakes—at least a meter high—in a corral in your garden. They should be separated by three or four centimeters—no more. Place some açaí fronds flat on the ground inside the circle. Tie a chicken by the leg inside the circle of stakes. You must leave her overnight. The pico da jaca will squeeze into the circle and swallow the chicken, but he'll be unable to squeeze back through the stakes because of his swollen belly. His natural instinct will be to hide under the açaí fronds and go to sleep. In the morning you can chop his head off.

"Now, Joaquím, listen carefully," Dona Cabocla admonished. "Picos da jacas always live in pairs, man and wife. And so you must catch the second one too. It will be very angry and will be seeking revenge. You must not let your children play in the rocinha until you have killed it, too."

Within a week Joaquím had killed two picos da jacas, each easily two meters long, and his family returned home.

Dona Cabocla and I became friends that morning, and I made a point of visiting her whenever I returned to Acre, whether or not I had business at Santa Luzia. She thought I was wildly impractical—a little addled, really—coming all the way to that most remote corner of Brazil just to collect plants.

"I can understand why you collect leaves—they may have some utility, as medicine, perhaps. But why do you take a tape measure into the forest?" she once asked me. "What value are all those numbers? Do people actually *pay* you to do these crazy things?"

One afternoon years later, I casually mentioned to Dona Cabocla that I had just returned from Valparaiso.

She was startled. "There are many places named Valparaiso. Do you mean the one on the Rio Azul? *That* Valparaiso?"

"*Esse,*" I replied.

"Oh, *meu filho,* let me tell you the story of that river. You see, I used to live there."

Dona Rosa Correira de Souza was born on August 8, 1928, on the Rio Juruá at the Seringal Paraná dos Moura. Her parents, from Ceará, were refugees from the devastation of centuries of deforestation in eastern Brazil. As in Amazonia, the tropical forests of Ceará used to generate their own rainfall, and demolishing the forests had inevitably led to drought. The second decade of the twentieth century was especially dry. Thousands starved, and entire villages had to be abandoned. As a child, Dona Cabocla's mother had to walk ten kilometers to collect a tin of water from a creek, then stagger back with it on her head. Taking advantage of the desperate circumstances in Ceará, representatives of the seringais of Acre actively recruited rubber tappers and dispatched them, all expenses paid, on steamships to Cruzeiro do Sul. Of course, the workers were obliged to repay their passage once they started collecting latex. Dona Cabocla's parents married in Ceará and promptly signed on for Acre. "It wasn't a civil marriage," she said, "but it took place in a church, so there's a record of my father's name written somewhere so that others can see it."

At the Seringal Paraná dos Moura, the couple fell into the unavoidable pattern of indenture and debt, and they were never able to leave. By the time Dona Cabocla was born, just before the Great Depression, prices for rubber had collapsed and many seringais were abandoned. The seringueiros struck out on their own, carving out *rocinhas* from the forest. Many of the greenhorns starved to death or were killed. Those with jeito got by.

"My given name is Rosa," she told me, "but nobody called me that. You see, when I was a kid, I was strong and fit. I never got sick. So they called me Cabocla."

Dona Cabocla married the boy next door, Manoel Arlindo da Fonte, at thirteen and moved away from her father's hearth to the frontier of the Rio Azul. The young couple first tapped rubber at the Seringal Ponte Bela, three days upstream from the river's mouth. For a while during the early years of World War II, rubber once again commanded high prices, and the couple were able to accumulate enough capital to move to the more prosperous Seringal Valparaiso, near the mouth of the Azul.

And that is how it happened that Dona Cabocla and Manoel chopped out a clearing, built a house, planted rice, corn, beans, bananas, papayas, manioc, and the jambu tree that now sheds its purple petals on river and land, just a few meanders from where we have made our camp.

"We chose the spot because it was on a restinga of terra firme," she once explained, "above the water's reach but close to the river and the várzea. But most important, it had terra roxa, sandy yet rich. We knew that it was good land and that we could farm it without having to move on. I knew that we could survive there when the latex didn't run or when the prices for rubber fell.

"You see, *meu filho,*" she continued gravely, watching for my reaction, "we listened to the bees, the *arapuás.* They show you where the best soils are." Dona Cabocla was referring to the bothersome but stingless sweat bees, no bigger than a house fly, that range throughout the New World tropics.

"The bees?" I asked.

"Yes. You see, the arapuás don't molest you in places where the soils are rich. In all the years we lived there, we almost never saw arapuás. But in the terra firme just behind the ridge, they were awful. We knew that our rocinha wouldn't wear out like the rest of the terra firme.

"We could tell that Indians had lived there before us," she continued. "The ridge was covered with their signs, and this is another indication of good soils. When we first began clearing the land, we found pottery everywhere. Stone axes, too, and stone weights for fish nets. The Indians knew the best spots."

Dona Cabocla and Manoel's instincts were correct: the terras roxas are good for farming. They are middens of food, bones, urine, and excrement—the day-to-day residue of living—left by generations of human occupation.

Manoel cut four tortuous estradas at Valpariso, each five to ten kilometers long, threading together about sixty trees. He and Dona Cabocla fell into the inevitable routine. He tapped rubber trees for eight months of the year, leaving the hut at one in the morning and finishing just before noon, while back home she smoked the latex into bolas and raised the children.

Dona Cabocla gave birth to eight children in that hut. Six— Antonio, Francisco, Gârdie, Francisca, Cláudio, and Maura— survived. When the babies were five months old, she weaned them onto a farinha paste, which she placed on their gums. But manioc, although gut-filling and therefore satisfying, has no protein. Lack of protein denies children physical strength and robs their brains of the nutritional building blocks of intelligence. Knowing this, Dona Cabocla had the good sense to give them fish from the river. At first she used a hook and line. Later she learned to cast a net.

One afternoon while he was chopping weeds in the rocinha with a machete, Manoel reached down to roll away a log and was struck by a baby pico da jaca.

"It was only fifteen centimeters long, as red as the earth itself. It was so small." Dona Cabocla paused. "It bit Manoel just on the tip of his finger. He didn't see the snake until it was too late." A baby pico da jaca may appear as innocuous as a worm, but it has the same anticoagulant venom as an adult. This one must have hit an artery. Within two hours Manoel had hemorrhaged to death through his mouth, eyes, nose, gums—even his fingernails. He vomited more blood than Dona Cabocla thought he could have in his body.

Dona Cabocla buried her husband at the edge of the rocinha near the purple-petaled jambu tree. She couldn't return to her parents on the Rio Juruá. Her family's demands would have pushed them all to starvation. Alone in the wilderness with six children, saddled with her husband's debt to the mateiro, she had little choice but to become a seringueira herself. For the next six years she eked out a living on a frontier where there was no comfort, no friend, no law.

Every morning at five or six, she left her house to tap rubber. A man would have left earlier to catch the time of maximum sap flow, but Dona Cabocla had to get breakfast for her kids.

"She worked very hard to sustain us," Gârdie, her oldest daughter, once told me. "But she couldn't have done it without my brothers." Antonio and Francisco, both twelve when their father died, took on the responsibilities of men: fishing, hunting, tapping rubber, and smoking the latex into bolas. "We weren't allowed to sell her rubber to a regatão," Gârdie continued. "We had no choice but to cooperate with the mateiro."

Dona Cabocla was able to pay off the aviação in six years. When she finally managed to move her family to Cruzeiro do Sul, she couldn't write her own name, but she was determined that her children would have an education. Dona Cabocla took any job she could get, no matter how menial, often working as a maid and nanny for the children of wealthy families. Soon Francisco joined the army and Antonio learned to operate a bulldozer on a construction gang building the Transamazônica. Both had steady incomes and began sending money home. Antonio learned about the government's plans to develop Santa Luzia while he was working on the highway; finding the idea of getting title to his own plot of land irresistible, he moved there with his mother and Francisco in 1982.

All of Dona Cabocla's children became comfortably ensconced in the middle class. Gârdie works for the government; Francisca and Maura are nurses. They have satellite TV.

I once asked Dona Cabocla how she got through those years on the Rio Azul.

"Oh," she replied with a shrug. "It is life. It is life."

I asked whether she thought she had been unlucky.

"Of course not, *meu filho*. I had good fortune. Am I not alive? Are not my babies living?"

A few years after my visit to Santa Luzia, the director there contracted cerebral malaria and died within forty-eight hours. The government was unable to recruit a successor. By 1988 Santa Luzia had suffered an epidemic of rabies, vectored by vampire bats. Four people and innumerable cattle died of the disease in the course of a few weeks. During each of the next three winters, falciparum malaria again swept into the settlement. It always arrived during the rainy season, when the road to Cruzeiro was washed out, so it was impossible to get treatment. Many of the original settlers lost family members to the disease. Things began to fall apart. The community truck was stolen. The government neglected to replace the water pump, and the onions died in the fields. The settlers went feral, resorting to subsistence farming, hunting, and fishing.

Santa Luzia and the other agrovilas were an economic boondoggle. The fundamental mistake was building the Transamazônica, the pilgrim highway that lurched across the sterile terra firme forests of Acre, indifferent to the life-giving várzea and the rivers. It was assumed that this celebrated road—an engineering scheme as monumental as the pyramids or the Great Wall of China—would always be there, and that the agrovilas would prosper. But the disintegration of the road forced the settlers to hew a living from the terra firme forest. Santa Luzia was doomed. The pioneers were stranded by their inexperience; they lacked the skills to survive in Amazonia and had no resources to return to the festering cities of the south.

On my final visit to Santa Luzia, in 1992, I rented a six-ton diesel truck, one of those indolent vehicles that had been stranded in Cruzeiro since the Transamazonica closed. The road to Santa Luzia, fifty-eight kilometers away, had become nearly impassable even during the dry season, and during the winter wet season, the residents had no way to escape except by canoe or by walking. Even though my visit was at the end of the summer dry season, our truck was the only vehicle to have made the journey over the ruined road that year.

The journey took all morning. The forest was gone, replaced by a chewed green *campo* of cattle pastures and a few emergent yellow crowns of pau d'arco. The igarapés, which once had coursed from the Jurúa and run full with fish, were stagnant and silted up. The two wooden bridges that crossed them were rotten. As we drove past the abandoned farms along the Transamazonica, the driver matter-of-factly described people's misfortunes: "This *morador* died of malaria, that one of a snake bite, another of malaria, of *gripe*. This man's cattle contracted rabies and he bailed out." Only the poorest, without the resources to move on, remained.

In Santa Luzia, abandoned by the government, chickens ranged through the ruined buildings. There were no funds to fix the water pump; the colonists had to walk half a kilometer down the road to the igarapé to bathe and get drinking water. Nobody dared fetch water after dark because jaguars roamed the periphery of the settlement. The administrative center was still there, but the air conditioning was broken and the bureaucrats had been replaced by a team of public health workers, brought in to deal with the malaria emergency. On the tables were laid out glass slides with blood smears. The workers told me that 70 percent of the settlers were infected. Everybody seemed idle, to be waiting for the end of this asinine enterprise. People were dying without remedy. The

ght kilometers from Cruzeiro might as well have been a
sand.

In contrast, Dona Cabocla's homestead was as verdant as I remembered, with healthy animals behind sensible hedges. The dirt
floor in front of her house was swept clean. She sat in her hammock, soothing a puppy that had the circular lesion of a vampire
bat bite on its flank.

"Oh, *meu filho!*" she exclaimed. "Why have you come back? Are
you here to take more crazy measurements?"

Her straight black hair had turned a bit grayer, but she was still
as stern and strong and bright as any person I have ever known. I
commented on how healthy her garden looked.

"You see, I have learned to make my own terra roxa from kitchen
scraps and chicken manure," she explained. "These terra firme soils
can be generous, but you have to give them life."

Dona Cabocla, ever pragmatic, had learned a new enterprise.
She had planted a field with coffee, pupunha palms, and guaraná,
a native vine that yields a caffeine-rich seed. Guaraná had become
all the rage in Japan, where it was regarded as a health food. Japanese merchants would visit Santa Luzia and buy Dona Cabocla's
entire crop on futures contracts.

"If the director had thought of guaraná instead of onions," she
sighed, "if he had understood terras roxas, then Santa Luzia might
have become prosperous."

Even before our arrival, the news that a truck was coming had
spread. A mother brought me a limp boy, begging me to take him
to the military hospital in Cruzeiro. I wasn't certain he would survive the journey, but of course I had to agree. Within an hour the
truck was loaded with refugees, its bed packed with people, dogs,
pigs, and the sundry items of entire households folded inside hammocks and tarps. I had planned to spend three weeks at Santa Luzia

to study the dynamics of my forest transect there. But I hadn't anticipated the outbreak of malaria. Worse, the rainy season was imminent, and there was a chance that our truck would be the last one out before the road washed out. Had I been alone, I might have returned, but I had three students with me, and I didn't want to expose them to unnecessary danger; it was, after all, my research, not theirs. So we left, our task unfinished. As I sat with the refugees on top of sacks of farinha, I wondered how they would make a living in a city that had no jobs, no reason for its existence.

"Why don't you come with us and join your family?" I asked Dona Cabocla before we departed.

"This is the only place on Earth that has ever belonged to *me*," she replied, patting her heart. "I love the solitude. And the trees . . . they're like my old friends. I was born in this forest and, God willing, I shall die in it."

"But there are dangers here—malaria, jaguars, loneliness," I replied. "Don't you want to live where it's safe?"

"No. Those things don't matter much to me anymore." Dona Cabocla sighed. "You see, when you become old, the world sort of loses its grip on you. I don't fear this place. What on Earth could still happen to me?"

ARITO

THE RIVER

DIONESIO

DONA CABOCLA

DONA AUSIRA

REFLECTIONS

CHIQUINHO AND SONS

TARZAN

ON THE LINE

Awake, O north wind; and come, thou south;
blow upon my garden,
that the spices thereof may flow out.

—Song of Songs

NOW WE HAVE COME to study that same terra firme where Dona Cabocla once tapped rubber. This is the gloomy, cathedral-like tropical rainforest of the popular imagination, with its escarpment of trees and shrubs and its vines that reach sixty meters toward the sunlight; except for light gaps created by fallen trees, the forest floor is deeply shaded and cool.

The stature and extravagant diversity of the terra firme forest make it easy to believe that the soils are brimming with nutrients. This belief convinced the designers of the Transamazonica that the terra firme could support vast croplands, so they lured thousands of immigrants from Brazil's impoverished east coast to the Amazon to become farmers. But they were deceived. In fact, the lowland terra firme is a place of senile, stingy soils leached of nutrients by millions of rainy seasons. As Santa Luzia proved, it is a place of hunger and nomads.

Where, then, are all the tons of minerals required to build this cathedral forest, if not in the soil? The answer is in the living biomass of the forest, in the plants and animals themselves. In temperate climates, where the land often freezes hard in the winter, the forces of decomposition can't keep up with productivity, and humus accumulates. This is especially true of prairies, the grasslands that are the breadbaskets of the world. Some tallgrass prairies of the American Midwest, for example, have topsoil that is three meters deep. In the wet tropics, by contrast, the decomposing agents—fungi and bacteria—are active year-round. Fallen leaves, twigs, and wood are rapidly broken down and their constituent nutrients returned to the living mantle of vegetation. There isn't enough time for much humus—let alone topsoil—to accumulate.

The cover of organic matter on the forest floor is usually less than a centimeter—perhaps only a few leaves—thick. If I scuff the earth, my boot easily perforates the humus layer and reveals a fibrous web of fine roots over a slick substrate of clay. The etiolated roots seem woven together, but they originate from any number of trees. This points to the underlying reason for the impoverishment of the terra firme soils. Not only does decomposition take place rapidly, but the released minerals are absorbed by the tree roots before they can permeate the soil. The trees find little advantage in sinking deep tap roots into these sterile soils. Instead, the roots lurk close to the surface, ready to intercept the falling litter. Their shallowness makes the terra firme forest sort of flat-footed and vulnerable to toppling—a characteristic that leads to increased heterogeneity.

The tropical forest floor, therefore, is a fabric of interlocking, warring roots. But the roots cannot absorb the nutrients on their own. Most of the trees enlist the help of fungi known as endomycorrhizae, which grow on top of or inside of the roots or intertwine with their root hairs. The mycorrhizae give the root mass a pale, fuzzy appearance and a distinct mushroom smell. Unencumbered by the lignin and cellulose of their hosts, the soft-bodied mycorrhizae assault the leaf litter, launching mycelia—threadlike filaments one cell wide—into the nutrient lode faster than any tree root could. A leaf is digested to a folded doily of white veins in a few days. A log becomes spongy in a month and is gone in a year.

This is a forest of nutrient avarice. Even the rain, graced with dust blown across the Atlantic from Africa, is worth mining; the leaves of many terra firme tree species scavenge nutrients from the rainwater before it hits the earth. (The dusty dry season in West Africa coincides with the rainy season in western Amazonia.) The result is that the terra firme forest is a near-perfect nutrient filter, and the igarapés that drain it run with water as pure as if it were distilled.

Yet in these sterile soils the wonderful panoply of earthly species has reached its zenith of diversity. The number of species in the terra firme, of just about every type of plant and animal, is greater by one or several orders of magnitude than in any other part of the planet. For example, the total number of tree species in Amazonia may be as high as 30,000. Inventories of trees conducted on small plots in the western Amazon have shown as many as 300 species per hectare. That's half as many tree species as occur in all of North America; four times as many as in Great Britain.

The purpose of our research would appear simple enough: to conduct an inventory of all the trees on an elongated quadrant—ten meters wide, starting near Dona Cabocla's jambu tree and running for two kilometers straight into the terra firme, over two ridges, into two swales and across an igarapé—a total of two hectares. We begin by laying out the transect, placing stakes every twenty-five meters and aligning them with a sighting compass—a frustrating job because the foliage deprives us of a clear line of sight. It would be far easier if we could cut a trail and clear away the understory, but that would preclude conducting future studies of how the forest changes, of its dynamics.

Then the drudgery begins. The days run together, the weeks become interminable. I had forgotten the consuming exhaustion of these expeditions. The team plods along the transect a decimeter at a time, like a plague of leaf-cutter ants. We somehow manage to map and measure the diameter, height, and crown of every tree greater than ten centimeters in diameter in the two hectares. That's a numbing total of 1,359 individual trees, and from each one we collect a voucher specimen—a small branch, some leaves, and, if we're lucky, some flowers or fruits. The mateiros clip the specimens with shears attached to a long aluminum pruning pole; if necessary, they climb the tree using a *peconha*, a canvas strap that binds their feet and enables them to shimmy up small trunks. This is where they earn their bread and butter. It's a dangerous job—they have to skirt wasps' nests and eyelash vipers—but they accomplish it with breathtaking élan. The mateiros are athletes, aerialists working without safety nets. They briefly become weightless beings, entering a canopy that is off-limits to most humans. I have watched them walk across the sky on broad boughs, through arboreal gardens festooned with bromeliads and orchids a meter high. They enter a realm that I, land-bound and envious, will never know.

It might seem idyllic to be an observant, contemplative naturalist methodically examining each life form in the solace of the vast wilderness. But our research has all the serenity of a factory stamping out garbage can lids. The emergent trees, fifty or sixty meters tall, are beyond the reach even of the mateiros, and in some cases we have to take our leaf samples with a shotgun. Every few minutes a shotgun blast pulverizes the tranquility, and the forest smells of cordite. The *tropeiro* birds shriek after every blast, evidently interpreting the alien sound as a territorial declaration, but no other animal has the nerve to be near us, and for most of the day the forest around us is evacuated. Now, after another shotgun blast, a baby *macaco barrigudo*—a woolly monkey—temporarily separated from its troop, is plaintively *eolking* in the trees overhead. And the constant babble is irritating. The mateiros, ladyless and randy, sing ribald songs and banter in a language of anatomical precision. Portuguese, it seems, has unlimited innovative metaphors for genitals. Worse, for a week now, we have all had explosive diarrhea from eating an armadillo left a little too long in the sun, and are continually creeping off to the edge of the transect to nourish the dung beetles, shouting to the others "*Eu vou matar uma paca*" ("I'm going to kill a paca"), a code that ensures a bit of privacy. It's the only private time we have all day, and all night, long.

Dona Cabocla was right. How inept it seems to measure this forest, this confusion of fallen branches, trunks, lianas, where nothing is linear, with a metric tape. How can a patently one-dimensional tool describe something so polydimensional? The web of numbers we generate traps only a few bits and shards of information that hardly seems to represent this place at all. Yet slowly, as we transmute the forest into numbers, we see a pattern and process that could never be discerned without them. There is a concise elegance in the way our experiment seduces nature into revealing herself.

The rules of assembly are distilled from the chaos. The inchoate becomes defined.

Except for volcanic oozings in a few remote ocean canyons, the sun is the only source of energy available to life on this planet. All green plants reach for those photons generated only nine minutes before, squeezed from the interstices of hydrogen atoms as they were glommed together by the amazing gravitational forces inside that orb a million kilometers across. Here in the rainforest everything reaches upward, away from the lethal shade. All these trunks and boughs are designed for only one purpose: to loft leaves into the light. It is only sixty meters from the forest floor to the tops of the tallest trees, yet success or failure is measured in this modest distance, and this extravagant architecture, this leaf scaffold, provides all the niches and chinks, refuges and retreats, for every living thing.

Only 0.5 to 1.0 percent of the light that falls on the top of the closed canopy of the terra firme manages to percolate to the forest floor at midday. During the morning and afternoon, when the sunlight is angular and its passage from canopy to earth is longer, the amount is even less. Most of the light that reaches the ground has been intercepted by green leaves and therefore does not have its full spectrum; the red and yellow bands requisite for photosynthesis have been filtered out. As one walks on the forest floor, one's eyes perceive the green light that survives this winnowing, light that is useless to green plants. The light scavenging herbs and shrubs of the understory are able to switch on their photosynthetic factories almost instantaneously and to survive on the scraps of leftover light. Now and then a shaft of full-spectrum sunlight noses through an interstice in the canopy and slants through the tiers of vegetation as if were in the nave of a cathedral. These beams create ephemeral "sun flecks" on the forest floor, where for a few minutes the illumination is intense. But the passage of the sun across the vault of sky and the ever-shifting shapes of the wind-tousled tops of the trees ensure that each fleck moves on and that the suffocating gloom quickly returns.

Most of the saplings are the progeny of the canopy-dwellers, and are thus light-demanding species. They wait for a chance to grab the sunlight; otherwise they have no prospect of surviving. Some wait on the forest floor for years, their growth stilled, nearly dormant. In this impoverished environment a sapling has the resources to make only a few leaves, each one of which represents a significant portion of its reserves. To invest in a leaf in this place of low returns is an act of great hope as well as great risk. The loss of that leaf—to a browser, to the inopportune step of a tapir, or to a casual flick of Tarzan's machete—is a catastrophe from which the sapling may never recover.

Of course, nature always favors cheaters, and many plants take shortcuts to the light-soaked canopy: vines, bromeliads, orchids, mosses, lichens, and liverworts that grow on another plants. To be an epiphyte is to have a place in the sun without making an investment in infrastructure, to fly to the treetops in the gut of a bird. By stealing light, epiphytes are in every sense parasites of their host trees. And they are liabilities in other ways, too. A tree cluttered with a garden of waterlogged epiphytes may lose a bough or even fall because of the burden. And if that tree is woven to its neighbors by lianas and creepers, its fall may bring down the whole neighborhood. An epiphyte's fortune is therefore directly dependent on its host's. Much of an epiphyte's success is just luck: a vine clambering over a geriatric tree has an uncertain future; one growing on a virile young trunk has good prospects.

The same storms that rend the atmosphere above tear the forest below. Downdrafts, whirlwinds, tornadoes, and screaming gales sheared from Antarctic cyclones all batter the western Amazon. This angry air topples the giant trees, which because of their nutrient avarice are shallow-rooted. Every one of the billions of trees in this forest will fall some day, in the process demolishing its smaller neighbors. The forest is a billion disasters waiting to happen. A treefall may devastate as much as a hectare, though most are much smaller, less than a thousand square meters. From the vantage point of a soaring *urubu,* the forest hardly looks uniform; it is more like a moth-eaten tapestry of fallen trees and the gaps they have made.

Walking on the forest floor, one frequently encounters tangled corpses of tree and vine, bright zones where the sunlight sears the skin and eye. Everything is different in these light gaps. The soil bakes and crumbles. Hot winds creep in, weeding out some species and encouraging others (a phenomenon known as the edge effect). Mosses and liverworts dry and flake away, and soft-leaved light scavengers wither. Photophobic insects, spiders, scorpions, millipedes, and centipedes flee or take shelter under logs.

But the light gaps are also places of regeneration. Many saplings have been waiting for years for this moment. Without the strangling shade, a leaf is no longer just a costly investment. Now the hopeful little trees invest in strength and, for the few months of unlimited sunlight, lurch toward the sky. They will inherit the void and close the canopy, and darkness will return until the next catastrophe.

Other light-gap trees are carpetbaggers, opportunists that exploit the momentary light but make no long-term commitment.

The cecropias, for example, live short, fast lives, investing little in woodiness and strength; they grow at a phenomenal rate—several centimeters a day. On a quiet night after a rain, you can hear the trees stretching and extending their jointed stems. Ultimately, cecropias are poor competitors and must disperse their seeds to a new gap before the canopy heals. For this they use the services of night-flying bats, which delight in the thousands of sweet, sticky seeds produced by the cecropia catkins. Most of the seeds are destroyed by digestion, but a few survive, stuck to the bats' fur or passed in the feces. Like the cecropias, the bats are gap specialists, flying like guided missiles from one clearing to the next. They sow the gaps with the species on which they depend.

Within a few years the cecropias will die, but their ephemeral shade will have restored coolness and humidity to the forest floor, and the striplings of other species will have been launched on their trajectories toward the sun. Other trees wait in the dark understory, their fates hinging on chance and contingency, on the luck of the next lightning strike, freak storm, or treefall. But it will take decades—perhaps a century or two—before the gap in the canopy fully heals. Along the way uncountable permutations of light, humidity, and soil dryness will occur, each permutation creating a microhabitat suited to one or two of the thousands of species of trees, mammals, birds, reptiles, amphibians, and, especially, insects adapted to these fleeting architectures. The "final" species composition of the healed gap is pretty much a crapshoot, a function of the particular seeds and seedlings that were waiting hopefully when the dominant tree fell. The gaps, therefore, make the forest unpredictable, create randomness; no single parcel of closed terra firme forest is representative of the whole.

Recent studies conducted in Panama and Costa Rica have shown that treefall gaps may not really heal, that they may be long-lived, self-perpetuating phenomena. Researchers theorize that the crowns of trees growing near the edge of a light gap, finding the light irresistible, lean into the void, growing disproportionally on their sunny as opposed to their shaded side. They are responding to a physiological imperative that served them well as saplings: lean toward the life-giving light or die. But trees with asymmetrical canopies tend to topple more easily than balanced ones, and their lives may be cut short.

The effect of this phenomenon on forest structure is enormous. Gaps, once believed to eventually heal, may be long-lived phenomena that transcend the lives of the trees around them. A gap may be the oldest feature of a forest of ancients, a hole that has been consuming trees for thousands of years.

Traditionally, fire was used to subdue these Brobdignagian trees, and the local tribal people and Caboclos use slash-and-burn, or swidden, agriculture to eke out a living from these soils. They cut the trees down in April, at the beginning of the dry season, leave them to dry all summer long, then burn them in September or early October, just before the rains arrive. After the burn, the soil, blessed by the nutrient-rich ash, is ideal for sowing crops. The soil nematodes and other pathogens have all been fricasseed—but so too have the mycorrhizae. The good times last only a year or two. After a few heavy bouts of rain, the nutrients wash away, leaving only the sterile red earth, which is soon compacted brick-hard by alternating sun and rain. The weeds and pests return. This is the deceit of the terra firme: beneficence followed by impoverishment. The local people have adapted to this reality by abandoning their plots before they fail, moving on and cutting new ones from virgin forest. Depending on the local circumstances, the progression from burning to bankruptcy takes between four and fifteen years. The terra firme forest is a place of nomads, of environmental refugees who continually shift from one swidden to another.

One could argue that the swiddens are nothing more than artificial light gaps and that the patchwork of small rocinhas mimics the natural mosaic of treefall gaps. Indeed, before Europeans brought iron to the New World, Amazonian swiddens were by necessity cut with stone tools, a daunting—if not impossible—task. It seems likely, therefore, that the native Amazonians took advantage of natural treefalls, rather than cutting intact forest. If that was the case, Pre-Columbian swiddens didn't mimic natural light gaps; they *were* light gaps. And as long as swidden farming was done on a small scale, within the dispersal range of seeds and propagules from the primary forest, the abandoned swiddens would eventually recover their full range of species.

But today's small-scale terra firme farmers have the tools to destroy the forest quickly and easily with chain saws and steel-headed axes. The swiddens are often of vast proportions, thousands of hectares, well beyond the reach of the seed bank, and their capacity to revert to forest is lost. They may never recover their diversity.

Our present understanding of the gap mosaic and its contribution to the environmental heterogeneity of tropical forests was a long time coming. Until a half-century ago, ecologists subscribed to the dogma that the exuberant diversity of the tropics was the result of long-term stability. This idea was championed by Alfred Russel Wallace (among others), who derived the theory of evolution by means of natural selection simultaneously with, yet independently of, Charles Darwin. Wallace, who learned natural history as a young man while traveling in South America, wrote of the Amazon forests:

> In the equable equatorial zone there is no such struggle against the climate. Every form of vegetation has become alike adapted to its genial heat and ample moisture, which has probably changed little even throughout geological periods; and the never ceasing struggle for existence between the various species in the same area has resulted in a nice balance of organic forces, which gives the advantage, now to one, now to another species, and prevents any one type of vegetation from monopolizing territory to the exclusion of the rest. The same general causes have led to the filling up of every place in nature with some specially adapted form. Thus we find a forest of smaller trees adapted to grow in the shade of the greater trees. Thus we find every tree supporting numerous other forms of vegetation, and some so crowded with epiphytes of various kinds that their forks and horizontal branches are veritable gardens.

The post-Newtonian world of Wallace and the other Victorian naturalists was unrelentingly mechanical. The Victorians assumed that, just as the planets cycled predictably around the sun and the stars voyaged across the heavens along fixed trajectories, every organism here on Earth was exquisitely and precisely knitted into the tapestry of living things. Each species was a sprocket in the great machinery of the ecosystem, with its own vital role. Each was a good citizen that put up good fences to separate it from the domain of its neighbor; each species was modeled to fit in the tightest possible adaptive hyperspace. It was a Panglossian paradigm: every species dwelled in its own best possible world, meticulously fitted into its particular time and space—its niche—in the physical world. That many of these dimensions were invisible to humans was simply a measure of our wimpy powers of perception.

This world view was compatible with the theories of both the creationists and the uppity evolutionists. It was irrelevant whether a precisely wrought species was the result of God's grand design or of evolution. God, after all, could be no less than a master crafts-man. And the process of natural selection, through painstaking trial and error, could result only in perfection, or something close to it.

The Panglossian perspective was reinforced by the notion that the tropics were imperturbable, the epitome of evolutionary sta-bility. After all, were not uniform temperatures, reliable rains, and abundant light emblematic of tropical forests? Was it not true that the Amazon had been static for 50 million years? Certainly there had been abundant time for every species to become a good neigh-bor so that the panoply of resources was shared amicably. This sce-nario depended on a certain economy, as well: no resources were wasted, every nutrient and photon was utilized to the fullest extent. After uncountable trials and errors, Earth and her children were liv-ing in perfect harmony.

But we now know that Amazonia has been anything but uniform either spatially or temporally, on either a vast or a small scale. Rather than a place of stasis, the Amazon has been repeatedly disrupted. For example, during the past 2.5 million years, polar ice caps have advanced and withdrawn approximately every 100,000 years. To-day we live in one of the interglacial periods, but during the last Ice Age (known as the Wisconsin in North America, the Würm in Eu-rope), glaciers more than two kilometers deep extended as far south as Iowa and Illinois in North America and completely smothered the British Isles and much of Siberia. These sheets of ice locked up vast quantities of water. At its peak 18,000 years ago, the Wiscon-sin Ice Age sequestered about 20 percent of Earth's freely circulat-ing water. Sea levels all over the world dropped by about 120 meters, and the moist equatorial forests in Africa and South America were invaded by fingers of dry savanna. The Amazon forest was probably broken into isolated patches that didn't dry out because of elevation or local climate, or it retreated to the river courses. Subpopulations of plants and animals, descended from common ancestral populations, became reproductively isolated from each other, then merged again after the Ice Ages ended. These repeated cycles of separation and consolidation may have created a "species pump," meaning that sibling populations in the isolated islands followed separate evolutionary tracks, differentiating from their common ancestor and becoming new species (or subspecies). The rate of speciation depended on the particular genetic characteristics of the founding population and on the particular selective pressures in each refugium.

Moreover, the creation of new taxa during the Pleistocene Ice Ages was potentially exponential, given that each Ice Age may have created fifteen to twenty separate refugia, each giving rise to a sibling species and that each of these new species, in turn, could have given rise to fifteen to twenty more during the next cycle of glaciation. Obviously, most of the new species didn't survive. Many would have hybridized with their siblings and blended once again into a single species. Others, requiring the same resources as their closely related siblings, competed unsuccessfully and were driven to extinction.

The process of competitive exclusion of sibling species is still going on in Amazonia today, and I expect that it will lead to a slow decline in the diversity of species over the millennia. However, in this dynamic forest, buffeted by frequent storms and treefalls, an individual tree is unlikely to achieve its full potential life span—and therefore to fully express its competitive edge against its neighbors—before being clobbered by a tipsy neighbor. In this ever-recovering place constantly rubbed down by the wind, the trees are a proletariat of short-lived survivors. No one species, even using all of the wiles sewn into its genes, can achieve dominance over the others.

Is the frequency of treefalls related to the species richness of Amazonia? The Amazon forest may provide a natural experiment that could answer that provocative question. Botanical inventories, like the ones we're making on the Rios Azul and Moa, have revealed that the forests of the western Amazon have more woody plant species than those of the middle or eastern Amazon. The inventories reveal, in fact, that the forests of western Acre—the ones we're studying on this expedition—are among the most diverse on the planet. Also western Amazonia has more rainfall than areas to the east. Several ecologists have linked these observations and postulated that rainfall is somehow related to the high species diversity. One can imagine, for example, that the magnitude and long duration of the rains in these western terra firme forests soften the slippery clay soils, resulting in more frequent treefalls. An elegant explanation, but I think there's more to the story.

The answer came to me on the night wind. One evening at Valparaiso, we were struck by a *friagem,* a persistent, unsettling, cold wind that stilled even the chorus of insects. The day before had been clear and hot, giving us no clue of a change in weather. But by dawn the forest was swaying voluptuously overhead. Shallow of root and poorly anchored, the trees eased back and forth as if they were standing on one broad foot; the two that I had strung my hammock to were gently rocking it.

Three or four friagems occur each summer in western Amazonia, lasting only a day or two. In 1854 Henry Walter Bates (who journeyed with Wallace in the eastern Amazon, after which he continued alone on the western Solimões), described a friagem:

One day a smart squall gave us a good lift onward; it came with a cold, fine, driving rain, which enveloped the desolate landscape as with a mist: the forest swayed and roared with the force of the gale, and flocks of birds were driven about in alarm over the tree tops . . . a most surprising circumstance in this otherwise uniformly sweltering climate. This is caused by the continuance of a cold wind, which blows from the south over the humid forests . . . The temperature is so much lowered that fishes die in the river Teffé, and are cast in considerable quantities on its shores . . . The inhabitants all suffer much from the cold, many of them wrapping themselves up in the warmest clothing they can get (blankets are here unknown) and shutting themselves indoors with a charcoal fire lighted.

The energy of the friagem surprised me. This was no fast-moving thunderstorm. All morning we heard branches snapping, and occasionally a reverberating treefall. Flurries of leaves turned in the air above our camp. None of the mateiros would venture into the woods for fear of being struck by a falling limb, so we spent the day in camp. It was a bit of a relief, really, because we were worn out by our weeks of travail on the line. What a luxury to lie in the hammock past dawn and listen to the rain-pelted leaves and the wind-shuffled canopy surging overhead, as if we were at the bottom of a shallow sea, looking up at the waves.

We know today that friagems are errant gyres of frigid air that rip north from Antarctica, push their way over the Andes, and slither into the Amazon valley. They are most violent near the Andes, and as they slide east, they rapidly dissipate. I wonder: in this ever-youthful forest, where no tree can grow to its full dimensions and maturity isn't an option, could the friagems—and the gaps they create—account for the unparalleled diversity? Could they explain the decline in diversity from western to eastern Amazonia?

Studying the patterns of the Pleistocene has taught us that destruction can lead to genesis, and the friagems teach us that chaos may nurture that diversity. Both concepts are counterintuitive to the perspectives of the Victorian naturalists who ventured in this forest more than a century before me. They knew little of the Ice Ages and less of Antarctic weather, but in this age of enlightenment, we have learned that Earth's continents and seas are configured in such a way that chaos—the life-giving forge of diversity—is the norm.

By late morning each of us is haloed by a swarm of arapuá bees that clamber over our faces and crawl into our nostrils, mouths, and ears, where they buzz and rattle, sipping beads of sweat, saliva, and mucus. They are our constant daylight companions. If you leave them alone, the arapuá soon depart, sated with excreted salt, but if you swat or accidentally crush one, she releases a pungent floral aroma that enrages her suicidal sisters. Although the arapuás are stingless, their sharp mandibles pinch the face and snip hair, eyebrows, and beard. I'm shedding onto my notebook as I write these words.

Dona Cabocla taught me to listen to the bees because they reveal the most important lesson of survival in this forest.

We are attractive to the arapuá because they are nutrient-starved; there are no salt licks here. These soils, far from the terra roxa of the river bank, lack the phosphorus, potassium, calcium, and sodium needed for life, and the insects are voracious for those minerals. In this counterfeit paradise, the arapuá regard us as irresistible sacks of goodies. Each of us is a nutrient bonanza, succulent and strange.

The arapuá aren't the only arthropods that crave us. The biting *mutuca* flies are particularly aggressive, harassing us day and night. Their recurved wings, faceted like panels of a stained glass window, enable them to maneuver adroitly in the closed forest. But they are

incautious, perhaps a bit drunk with all this easy blood, and therefore easy to slap; they die with a satisfying crunch. Yesterday the
mucuims—burrowing mites that set up shop under your skin—invaded my waist and stomach, raising angry welts. They are living
inside me now, breathing through little chitin snorkels that feel like
embedded slivers of glass, borrowing my warmth and energy for a
few days until they abandon me. My body has tried to reject them,
attacking them with white cells and histamine humors, but they are
as indifferent to my defenses as a tin roof to the rain. Today I made
the mistake of resting in a patch of flattened grass—probably where
a deer or agouti had slept—and picked up uncountable juvenile seed
ticks, each the size of a pinhead. I scraped eighty off one leg, but I
didn't get them all. They leave little squares of macerated skin that
will seep for several days before scabbing over.

To ward off pests I have tried keeping mothballs in each of the
four pockets in my pants. Every morning before I leave camp, I sat-
urate my shoes, socks, pants cuffs, and waist with repellent. It erodes
the rubber binding of my notebooks and dissolves the lacquer on
my pencils, causing them to stick to my fingers. But these measures
do little to deter the parasites.

Listen to the bees. I have become the dinner of every bee, chigger, mosquito, tick, and biting fly that has evolved to suck the juices
of the creatures that live in this forest. My nutrients are flying and
buzzing everywhere in this forest and will soon be passed on in
humble, hopeful eggs. The calcium atom that yesterday resided in
the capillary wall of my finger is now depolarizing a neuron in the
brain of a mosquito, enabling her to avoid my slap.

At four every afternoon we return to camp, limp with exhaustion
and soggy with sweat. I am afflicted with a dull heat-headache and

stippled with punctures. My left index finger has two fang marks, indurated and angry, and is swollen to twice its normal size. I didn't even notice the spider; the color of lichen, it was probably running up a trunk when I put my hand on it. All these sores are eased by a bath in the river, but the headache persists. Now I sit in the soothing water, eyes closed, allowing the cold current from the Serra Divisor to massage me. After a while the top minnows find me and, tentatively at first, begin to nibble away the seed ticks embedded in my skin. They learn quickly that I am a grateful host; I am being groomed by a swarm of fish, by a thousand sharp, blessed little teeth.

THE NAMING OF PARTS

What's in a name? that which we call a rose
By any other name would smell as sweet.

—William Shakespeare

I AWAKEN before first light to a blasphemy of *guaribas*—red howler monkeys—their voices reverberating across the water, punching through the mist. Among all Earthlings, no other has a voice even remotely like theirs. Guaribas are the sequoias of animal vocalization, producing one of the most fascinating, yet discomfiting, sounds in nature. First one hears a few grunts from the dominant males, then some short barks. Then all hell breaks loose: an unearthly cacophony of thirty or forty voices, each amplified by a larynx the size of a fist. I can discern individuals in the bedlam: high-pitched shrills, growls, gurgles, snarls, bellows, gibbers. The sounds are so loud that when I close my eyes I see colors. I have long anticipated this moment of attempting to bind the voices of guaribas into words in my own language. But no turn of phrase suffices for this utterly ineffable event. My notebook is full of deletions, rubbings, lines scratched out. Now, perhaps, I have it: the sound of a demonic wind screaming through the rigging of a ship?

Why do howlers bother to make all this racket? It certainly must make them conspicuous to the harpy eagles that course over the treetops. As with so many of the peculiarities of this forest, the answer is related to the pervasive nutrient impoverishment. The heat and sun create enormous primary productivity here, but little of it is of use to animals (or, as we know, to humans). And howlers have a harder time than most: they are the only New World monkeys adapted to a diet of leaves, which are infamously low in nutrients and, worse, often laced with poisonous alkaloids and other toxins. The howlers cope with all this roughage and poison by being epicures, specializing on only the most digestible of the thousands of species of trees in this forest and browsing on only the youngest, most succulent leaves, before they have accumulated toxins. They are obligatory nomads, moving on before they run out of leaves in a particular spot. A troop of howlers, therefore, requires an enormous territory—forty or fifty square kilometers—which it defends with the most economical of all tools, one that can travel for kilometers yet costs next to nothing to make: sound. They vocally mark their territories for the same reason a jaguar sprays a tree with his urine: to keep others of their kind at bay without engaging in costly combat.

The guaribas have stopped shouting now and, as is their imperative, are clearing out. They won't return until these trees have refoliated. For a moment the newly risen sun refracts in the alarming red mane of a large male. He disappears, but his shifting shadow wiggles over the broad leaves of an *inajá* palm, and his musk—a mix of BO, urine, and feces—will linger in the still air all day long.

A low, gray shield of clouds hugs the river and forest. The air is stifling and still day and night. Arito says it is always this way after the first spats of rain; every living thing is flummoxed by the humidity, but there is not yet enough moisture for renaissance. The forest seems to be expectant, its trillions of leaves wilted; billions of animals are tucked safely into the long sleep of estivation, longing for the first soaking. For most of the day the forest is bird-silent, save for the Doppler shrieks of the tropeiros and the squawks of an occasional flock of thirty or forty mealy parrots. They forage through the canopy like a raucous group of schoolchildren, embracing it with claw, prehensile bill, and voice. In tongues that mimic the sounds of almost every other creature in the forest, they guard their mobile acoustical domain.

The transect has become our dominion, its limits our horizon. For a month we have been prospecting its course, learning its every gap and nuance, examining every tree, bee hole and log, mapping every darkling understory plant that scrounges a living from the light flecks and every tree that seeks the light above. Now the long-anticipated rains are transforming bits and pieces of the river and forest, and even the passage of a day brings change. A thorny *Zanthoxylum,* barren yesterday, unfolds pale green leaves today. The *Geonoma* palms are unfurling new leaves laced with shocking-pink anthocyanins, and the flat-winged seeds, as big as saucers, of *Aspidosoperma* are pushing bright new cotyledons through the leaf litter.

Zé Brejo has cut a trail set off from the research transect by about ten meters, and this has become our route to work. The transect may be compass-straight, but Zé Brejo's trail, avoiding obstacles and investigating interesting nooks, wanders through the forest. At kilometer one, a massive treefall; it must have knocked down half a hectare of trees, including jauarí palm that is shedding uncountable spines onto the trail. Zé Brejo cuts a detour around the jumble of branches, vines, bromeliads, and orchids, and the path permanently adopts a new bend. We make another detour out of deference to a mother *beija-flor-de-banda-branca*—a versicolored emerald hummingbird—that has woven her thumb-sized nest into a twig at waist level. Our trail, distorted by these moments of death and maternity, tacks away from our destination and back again. Soon the reasons for these diversions will have rotted and vanished, and the caprices of our trail will be their only witness.

On the morning walk to work I like to stride ahead of the others in order to catch a glimpse of the startled animals. These visions are evanescent, and being second won't do: the backside of a tapir; a paca like an animate sun spot; a brown egg-eating snake rising like a cobra (even though it's just a bluff); an *Ameiva* lizard looking like animate mercury; a brown rain frog plastering his egg spume onto a vine over a rain puddle. Now a surprised *caranguejeira* spider the size of a supper plate. The tarantula stands her ground, erecting her abdomen like a cat in heat. The cognoscente of the forest recognize this as a warning, for her abdomen is far more dangerous than her fangs. It is cloaked in a down of brittle, urticating hairs that break off like shards of glass and can drift into an open eye, scarring the cornea. I cajole her onto the blade of my machete, where she stands on spindly haunches, tasting the air with her forelegs.

But there is also danger in going first. Ahead, hanging under the broad leaf of a *bananeira,* is the shaved-paper nest of several hundred acabas da noite. They are asleep now, their yellow abdomens pressed in orderly rows on the paper casque of their house, which looks like a custard apple. Bumping into a wasp nest is probably the greatest danger in any tropical forest. You can get a hundred stings before you know it. The comings and goings of diurnal wasps reveal the location of their nests, and even a casually observant person can avoid them. But because acabas do noite sleep during the day, their nests are easy to miss. Therefore one *never* strides past a rosette of low-slung palm leaves or a clump of bananeiras without first peering underneath. And listening. Ahead, the sisters of a nest of a particularly nasty wasp, the *acaba tatu,* are synchronously drumming their heads against dry paper . . . *tica, tica, tica* . . . an aposematic tattoo that tells us to stay away. The wasps don't want a confrontation any more than we do.

Now an arresting moment of color: a male Achilles morpho butterfly, swift and agile, is flying through the understory, darting over and under and among vines and trunks. The insides of his wings are mercury-blue-bright and each downstroke is a burst of light. Most poisonous insects of the forest are brightly colored, an advertisement that warns predators of their nastiness and therefore discourages them from attacking. But morphos are highly edible, as revealed by the countless morpho wings I've found scattered under the perches of flycatchers and jacamars. Why, then, do the morphos advertise? I'm convinced that it's not advertisement at all but mimicry—not the conventional cryptic mimicry of a piece of bark or a dried twig but of a leaf tumbling from the canopy to the ground, its facets turning flamboyantly in the shafts of sunlight. A falling leaf is one of the most commonplace motifs in this forest, an occurrence that every predatory bird must learn to ignore, lest it be aroused every time the wind shudders the canopy. Birds, therefore, have become conditioned not to notice the morphos.

A quarter-kilometer farther down the trail, a centenarian samaúma grows sixty meters tall, emerging above the canopy. Its trunk must be six meters in diameter, its flat buttresses five meters wide, tapering to undulating roots that cross the forest floor like congealed boa constrictors. The trunk of the samaúma is hollow, its side rent, revealing a dark interior. Long ago its heartwood was consumed by fungi and countless generations of beetle larvae, and now all of the weight of this giant is borne by a shell of bark and green cambium. At first this would seem to be a disadvantage, dooming the tree to topple in the next strong storm. In fact, I believe, the hollow trunk is adaptive. Peer inside, and you will find that the vaulted ramparts are home to several hundred dogfaced bats, hanging like furry fruits. Below them, roach nymphs, scorpions, and beetles snuggle in the spongy bat guano. I wonder: Does the tree invite the bats? After eating nitrogen-rich insects in the far-flung nooks of this forest, do they carry nutrients home as guano? Are the minerals and ammoniac nitrogen the edge that the sama-úma requires to spread its arms above all its neighbors and possess the sky? Is a hollow heart and shortened life the price this tree will pay for its moment in the sun?

The forest floor is crossed by narrow paths swept clean of vegetation and debris: the highways of leaf-cutting *saúva* ants. The saú-vas are all sisters, haploid daughters of the same queen. Collectively thay are a single organism with a million and a half feet, a half-million eyes, a quarter-million brains, all coordinated by the acid scent they spray on trails on the forest floor, dribble on tree trunks, and paint on leaves. It is as if scent were the sinews and tendons of a superorganism diffusely insinuated into every hidden nook of the forest, as if each ant were a bit of tissue with a mote of consciousness and an atom of free will.

This is a forest of ants. Their biomass is greater than that of any other type of animal here—birds, reptiles, mammals, even beetles. After a rain, the air becomes pungent with the slightly stingy, eye-watering smell of their formic acid–laced trails. My clothes and the damp pages of my notebook reek of formic acid; my hair is marinated in it. There are tens of thousands of leaf-cutters on the trail this morning, each hefting a shard of green leaf that weighs a hundred times more than it does, all marching determinedly to their subterranean nest somewhere nearby. Arthropod agriculturists, the leaf-cutters won't eat the leaves they are harvesting. How could they, with all the leaves' poisons, lignins, and tough cellulose? Instead, they masticate the leaves to produce a mulch on which they sow soft, edible fungi. Locked away in underworld chambers of precise temperature and humidity, the fungi do the hard work of digesting, breaking down all the inedible substances that the plants have worked so hard to make.

🐌

Yesterday on our walk to work we encountered a band of queixadas—white-lipped peccaries. Only the size of beagles, the peccaries have flat, scissorslike canines that administer death by slicing. They are regarded as the most dangerous mammals in Amazonia. The first thing we noticed was their musk, which put us all on edge. Each of us immediately selected a small tree that would be easy to climb. Soon we heard them, smartly clacking their jaws, and then a band of about fifty surrounded us, appearing as if from nowhere—boars, mothers, suckling babies—a vocal bunch all communicating with a curiously diverse vocabulary of low-frequency belly-grunts.

White-lipped peccaries are especially deadly because they have an instinct for public service, a sense of altruism that motivates every member of the band to defend the others to the death.

"If you shoot one," Arito explained, "the others will come to its rescue, nipping at your legs and, especially, your Achilles tendon. They know how to incapacitate a man; they know our weak places. Even being on a horse is no advantage, for they'll cut his heels, too. I once saw a jaguar that was killed by peccaries when it became greedy and refused to abandon its kill. It didn't have the good sense to climb a tree. A smart jaguar will always take its prey into the low branches."

So we climbed our trees and waited. The perch need be only a meter or so above the ground to be beyond the reach of even the biggest peccary. They nonchalantly foraged under us, snuffling tubers and grubs from the leaf litter, now and then squinting upward and snuffing the air. After a while they moved on. But we were spooked. All afternoon we were watchful and deer-fey.

At midday we choose a shady place, close to some convenient peccary-trees, for lunch. The parrots and tropeiros are silent; the peccaries are in their wallows, the *Ameiva* lizards have disappeared into the shading leaf litter. Only a few furtive nymphalid butterflies are active, tumbling in the hot air and tippling the red inflorescence of an *Ixora*. The arapuá bees are buzzing so loudly around my ears that I can hardly hear myself think. Lunch is a moment of respite, of contemplation, when the men—their mouths full—finally stop talking. Zé Brejo has shimmied up a *bacuri* tree, carrying his machete in his teeth, his back legs splayed like an iguana's; he tosses us a branch laden with orange-yellow fruits. Bacuri fruits are astringent, have a slight terpene taste, and are notably good at quenching thirst. Tarzan rolls a cigarette of coringa in a wrapper made from bark stripped from an *envira* tree. The good blue smoke is so rich that it repels the white-footed *Anopheles* mosquitoes. As I sit on a fallen log, writing, the day's heat is overwhelming. A few trickles of perspiration slide down my belly. The *pium* flies nibble my ankles when the wind dies. And the wind has died.

Every day at this hour, I try to describe my environment in detail, to write down what all my senses are saying. This act of forced transcription disciplines me to understand this place, to observe instead of slipping into hot, fitful sleep. Look: a broad, rapidly moving column of tiny ants, finding the spilled oil from a sardine tin, is rising like a column of living smoke over a log. Sardines are rich in all of the primordial, alien salts of the sea, and the ants are addled by this unexpected bounty. Most race upward, tracking the scent of their predecessors; a few, zigzagging like pinballs, run against the tide. Look: a hunting wasp, as blue as flame-burnished steel, her wing joints clattering as she flies, legs trailing, is patrolling the leaflets of a *Maximiliana* palm. When she alights, her shadow is backlit through the striated green leaflets. Her antennae tap a frond, tasting it. Now she ambushes a wolf spider, stings it, lugs its paralyzed body to the edge of the leaf, and launches herself, flying low and heavy . . . in fact, too heavy. She plops to the ground, drags the near-corpse up into a low shrub, and lifts off once more.

A *bico de argulha*—a jacamar—waits on a twig, stiletto-still, monitoring the insects that pass through a light fleck. Now it spots a darner laying her eggs in the flooded chalice of a bromeliad, dipping the tip of her abdomen into the still pool. After each bout of egg-laying the darner rises—using her fixed compound eyes to interpret the air for an enemy's movement—then dips again. But the jacamar darts into the shadows and attacks from darkness, clipping the insect in midair like a pair of tweezers. It carries the darner to a low bough and, with violent swats of its head, clubs her to pieces. The thuds are audible on the moss-covered branch. The darner's thorax crumples; her four clear wings twist in different directions. One wing falls to the forest floor, turning in the still air, intercepts a shaft of sunlight, and splinters it into a momentary rainbow.

I am contemplating a smattering of maggoty orange *abiu* fruits dropped from somewhere in the canopy. They are surrounded by a miasma of fruit flies, motes of life no bigger than a fleck of lint, which lay their eggs in the rotting stew. Zé Brejo calls the abiu *frutas do jabutí* and explains that all one has to do is wait beneath the mother tree for the tortoises to appear. And he's right. Today three jabutís, including a couple in amplexus, are gorging on the abiu, maggots and all. The male slides his long, hard tail under hers, straining and audibly groaning, while she eats demurely.

Every living atom of this forest has its moment, its particular occupancy of space and time. Take, for instance, the bed of fallen abiu fruits. To me they look rather uniform, although a few are discolored where they have been bitten by tortoises. But to a fruit fly, an abiu fruit is a terrain as complex and individual as a continent. The fruits release aromatic compounds that waft through the forest and attract the flies. Depending on how long ago it fell, each fruit is in a different stage of decomposition, colonized by a sequence of bacteria, fungi, and insects. The lives of these seemingly trivial little ecosystems have pattern and order. Studies of decomposing fruits of a species similar to the abiu have shown that a single fruit may support almost a billion individual yeasts of several dozen species over the several weeks it takes to rot. Most of the yeast species colonize sequentially, specializing on a particular combination of the ever-changing mix of sugars, alcohols, and acids of decomposition. The fruit-fly maggots don't feed on the sugar-laced fruits themselves, which are crafted to appease the sweet tooth of monkeys; lacking protein, they cannot provide the building blocks that the growing larvae need. Instead, the maggots fatten on the protein-rich yeasts. The adult fruit flies, their guts brimming with spores from their last supper, probably inoculate the abiu fruits with the yeast species they need. They sow their own gardens.

Such sequences of decomposition are not unique to abiu fruits. Similar successions in yeasts have been identified in rotting domestic tomatoes. But consider this: there may be twenty thousand species of fruiting trees in the Amazon, each with its own particular way of decomposing. No one could possibly identify all the types of fruit—let alone all the species of yeast—in this place. Decomposition is taking place all round me, among the abius at my feet and on the wild cacao pods by the river. I can smell the lovely yeasts, and the fruit flies, each bearing a bouquet of spores, enliven the still air. But I am incapable of appreciating this panoply. After all, I am only human.

How do I decipher pattern among the billions of light-seeking leaves, all hanging on to a tier in the canopy, a chink in the bark? How do I interpret this bedlam of sepal and petal, fruit and yeast, wing and scale? Grasping the parts of this forest is a bit like learning a language: you absorb a few syllables at a time, daunted by the knowledge that you can never finish the job. Learning to recognize one or two species is a good day's work. And if the species are the words in this new vocabulary, then the voucher specimens we collect for each individual tree are the syllables. A voucher is no more than a dried sample of foliage, flowers, or fruit. It is the irreducible element of our analysis of this forest's structure and diversity. Vouchers are a necessity in a region for which there are no field guides or comprehensive taxonomic keys, none of the handy conveniences that facilitate identification in the temperate north. Roger Tory Peterson never walked here. Our vouchers will be sent to taxonomists in herbaria, universities, and museums all over the world, where they will be compared with specimens already archived. If we're lucky, they will identify them to species. Our vouchers will provide proof to a skeptical scientific community of the diversity that we claim. Without those specimens, our conclusions would become no more than unreproducible opinions formed long ago in a place far away, a place perhaps forever destroyed.

Unfortunately, most of our vouchers are sterile; that is, they lack flowers, seeds, or fruits. Sterile vouchers, which are less than fully diagnostic, are the bane of taxonomists, who, following the precedent set by the Swedish botanist Carolus Linnaeus, rely on a plant's sex organs to construct their keys. Although Linnaeus did not understand evolutionary theory, he knew that a plant's reproductive parts, which are adapted to pollinators, dispersers, water, and wind, are more diagnostic of species than the vegetative parts. We know now that small changes in the morphology of a flower can reproductively isolate a lineage from its ancestors and lead to the genesis of a new species. And the adaptations of fruits and seeds—devised to stick to the fur of a mammal, pass through the gut of a fish, or bob in the flooded plain—are equally diagnostic. Therefore, taxonomists celebrate the sex life of plants; they are voyeurs who use a plant's reproductive organs for their analyses.

We fold each voucher specimen inside sheets of newspaper and bundle it tightly to take back to camp. The folios of newspaper are unsold runs of *O Jornada,* a scrappy Manaus daily. Every page is the same, and it's old news: hysterical screeching headlines: POLICE DESTROY BARRACKS AND SQUATTERS FLEE THE LAND; SEXUAL MANIAC KILLS WENCH WITH A KNIFE TO THE HEART; a photo of Fabio Jr. surrounded by lovesick teenyboppers. All day I plaster the limp specimens into Fabio's leering maw. This hay-baling would be tedious enough if we collected only one voucher from each plant, but we collect duplicates as well—eleven of each fertile one—a surplus that will be shared with other scientists. From Valparaiso alone we will have about 10,000 specimens, enough to fill up most of the space on *Fe em Deus.*

It is impossible to walk through this place where life covers every square centimeter without disturbing that life. The very act of measurement changes the experiment. The team tramples the understory, bends saplings, squishes footsteps into the soil, denying the patient recruits their place in the sun. In our passage we snap vines, cutting off hundreds of meters of living material in the canopy above. Even a casual blow of a machete has consequences: the loss of a single leaf could be fatal to a seedling that has only three. As they scale the trunks with peconhas, the mateiros bruise the bark, providing portals for invading fungi, grubs, and borers, thus affecting the tree's health, if not shortening its life. They often have to amputate an entire limb just to collect a few leaves, rendering valueless the tree's investment in that appendage and the mighty infrastructure necessary to project it into the sky. Dionesio slashes the bark of every tree with his machete, noting the color, layering, and striations of the cambium. He smells and tastes the sap and rubs it between his fingers to feel its viscosity and texture. His nose is as well trained as a jaguar's. From simple chemical clues he can usually identify the family and often the genus. The aluminum nails we pound into each tree to secure the metal identification tag create an angry inflammatory response: a lump of swollen and cracked bark that oozes sap. The scars of our research mark every one of our beloved trees. To know them we must injure them.

But the trail itself may be the most insidious result of our visit to this forest. As soon as we leave, it will become irresistible to hunters and gatherers. Moreover, a path marked by shiny aluminum tags allows hunters an easy return to camp on dark nights. Used repeatedly, our transect could well become a permanent conduit. The mortality and recruitment of trees will be forever changed. Fast-growing species will be favored, and the turnover rate of our small sample of the forest will appear to be accelerated.

In the evening, after the annoying arapuá bees and pium flies have stopped flying and we have gathered around the campfire for a big

supper and a few cachaças, we process the day's vouchers. It is a me-
thodical and contemplative process, very different from the haste
and ruckus of our work on the transect. The specimens are still pliable, and we arrange them to reveal the most information: folding a few leaves to show both sides, making sure that all the flower
parts are present, slicing the fruits so that they dry evenly. Then we
compress the vouchers between blotters and bake them overnight
over butane stoves. Properly dried, mounted, and protected from
fungi and insects, these herbarium specimens will last for hundreds
of years.

It is a magnificent journey to examine one of these specimens.
My science is to decipher how each species fits into the tapestry of
trees. What is the adaptive terrain of the tree's architecture? How
does it branch? How does it present its leaves to the life-giving sun?
How does its bark repel climbers? How do its flowers seduce bees,
butterflies, or ants? Each species has an array of gifts. But first we
have to assign every specimen to a morphocategory—a putative
species. We do this by spreading all the dried specimens on the
ground and sorting those that appear identical into piles. Each pile,
we presume, represents a species, even though we may not know
its name. Careful—the leaves of an individual tree may vary enor-
mously: its understory leaves are usually big and flat in order to
scavenge every last photon from the trickling light, but the canopy
leaves are economical and small in order to conserve water in the
wind-blown top. Sometimes I have to contemplate a specimen for
a long time—take a stroll, perhaps—before deciding whether it is
the same as or different from its neighbors.

I am now holding a voucher specimen from tree number 1,121
(of a total of 1,378), which grows 102.5 meters from the origin of the
transect. Its trunk is 17.5 centimeters in diameter; its height 12 meters. Clearly, this individual is in the genus *Inga* (one of sixty *Ingas*

in our sample). *Inga*s are easy to identify: all have an odd number of compound leaflets and most have at least one extrafloral nectary, which looks like a little mouth at the juncture of the first two leaflets. Some *Inga* species have full-lipped nectaries, others are thin; some nectaries are crystalline with sugar, others are empty at this time of the year. The nectaries nourish visiting ants, which in return repel predators and use their mandibles to clip away any overgrowing vines, mosses, lichens, and liverworts. The ants are easy to please, accepting the simple rewards of sugar and starch, which are far cheaper to manufacture, in this nutrient-starved place, than the nitrogen-rich alkaloids required to poison a browser.

There are about three hundred species of *Inga* in the New World tropics. About a third have appeared since the beginning of the Ice Ages 3.5 million years ago—a microsecond in Earth's lifetime, but long enough for them to have evolved explosively. Each species is a suite of genes shared by a diffuse population of individuals with a particular occupation of time and space. This time-place is known as a niche, and although "niche" is the central theoretical concept of ecology, it is intractably elusive because most of the dimensions of a niche are invisible to us. Armed with puny primate senses and only a rudimentary understanding of this environment, I can never fully grasp the niche of this *Inga*. Built into the genetic language of this plant is an understanding of solar physics, the wicking dry wind, and the rain of Sahelian dust. It bears the instructions for making leaflets, parenchyma, stomata, petioles, petiolules, rachises, nectaries, pigments, stipules, roots, root hairs, pipecolic acids—the list of structures could go on for pages. In this most complex of terrestrial biomes, the niche of an *Inga* also has an equally confounding biotic dimension. Small-flowering *Inga* species are pollinated by bees; large-flowering species by bats. The petals, bracts, sepals, stamens, pistils, ovaries, and other flower parts must be crafted to lure and satisfy these animals. And the niche has a gustatory dimension: all *Ingas* produce a succulent white tissue known as the sarcotesta that envelops its seeds. The sarcotesta is irresistibly sweet (perhaps nourishing, as well) to monkeys and is no doubt their reward for dispersing the seeds. Humans, being primates, share the same sweet tooth and take delight in eating *Inga* fruits throughout the range of the genus. How, I wonder, does the *Inga* know what I like?

After a few hours of attention to the minutiae of each specimen, Dionesio and I have concluded that the sixty individual *Ingas* constitute twenty-one species, all variations on a theme and all living in this small parcel of forest. We have classified most as morphocategories, but a few are recognizable as known species. The plant I am holding is an *Inga macrophylla*, a species that ranges from the Atlantic to the Pacific. Botanists use an arcane but expressive language to describe the traits that differentiate species, traits that, as well, partially define their ecological niches. The following formal description of *Inga macrophylla*, from the most recent technical monograph of the genus, published in 1999, by taxonomist Terry Pennington, tells us, in part, its niche:

> Young shoots 4-angled, pale lenticellate, pubescent to glabrous. Stipules 0.7–2 cm long, ovate, striate, pubescent or glabrous, persistent or not. Petiole (3.2–) 4.5–6.5 cm long, terete, pubescent or glabrous, rachis 10–17 cm long, winged (up to 2 cm wide), scattered pubescence, appendix to 1 cm long, linear, or absent. Foliar nectaries sessile or stalked, cup-shaped or unexpanded, 1–2.5 mm diam. Petiolule 2–4 mm long. Leaflets (2–)3–4 pairs, terminal pair 13–27 × 6–14 cm, broadly elliptic, apex narrowly attenuate or acute, base rounded; basal pair 7.8–17.5 × 4.5–8.8 cm, shape same as terminal pair; subglabrous above, scattered pubescence or glabrous below, frequently with minute red glandular hairs; venation eucamptodromous to brochidodromous; secondary veins 8–15 pairs, parallel or slightly convergent,

arcuate; intersecondaries moderate; tertiaries oblique. Inflorescence axillary, often clustered around the apex in the axils of the undeveloped leaves, solitary or paired, a congested spike; peduncle 2–8 cm long, pubescent to glabrous, floral rachis 1.8–6 cm long; bracts 1–3 cm long, lanceolate to ovate, striate, persistent; flowers sessile. Calyx closed in bud; tube 1.5–2.7 cm long, tubular, sometimes curved, striate, lobes 2–10 mm long; scattered hairs on lobes, otherwise glabrous. Corolla tube 3–6 cm long, lobes 3–6 mm long; pale villose. Stamens 65–100, staminal tube 3.4–7.2 cm long, 1.5–2 mm diam., equalling the corolla or exserted, free filaments 2.5–3.5 cm long. Ovary of 1(–2) carpels, glabrous, style exceeding the stamens, style head expanded, ovules c. 30. Legume 20–45 × 2.5–4.5 × 0.7–1.7 cm, quadrangular, straight or slightly curved, apex and base rounded or obtuse, faces with faint transverse venation, margins 0.7–1.7 thick, winged (to 6 mm); hirsute to glabrous.

The vocabulary of botanical anatomy may seem arcane to the novice, but it is finite. And that may be its principal virtue, for it allows the categorization of all this variety into a few recurring motifs. There are, after all, a limited number of ways to be an *Inga*. Pennington's description of *I. macrophylla* presents seventy-six differentiating characteristics, describing about as many dimensions of the species' niche, but these are only a small part of the whole. In conformity with the tradition established by Linnaeus, most of Pennington's elements pertain to the plant's sexual organs. Those elements are sufficient to differentiate it to species, but they represent only a small part of the thousands of dimensions that *I. macrophylla* occupies in this forest.

Pennington is the latest in a lineage of taxonomists who have classified the genus *Inga*. Since its discovery in 1806, the species currently known as *I. macrophylla* has had a variety of names and synonyms: *I. bracteosa, I. brachyptera, I. calcocephala, I. macrophylla* var. *ste-*

noptera, I. ouraphylla, I. alatocarpa, I. chorrerana, I. quadrangularis, Feuilleea bracteosa, F. brachyptera, F. macrophylla, and *Mimosa macrophylla.* All of these names appeared in the taxonomic literature at one time or another but were judged by Pennington to be synonymous with *I. macrophylla.* Because of its edible sarcotesta, *I. macrophylla* has become a beloved plant, widely recognized and cultivated (or at least encouraged) throughout Amazonia. It therefore has many indigenous names: *guaba, guabo* (in western Ecuador), *guambushi* (among the Shuar and Zamora-Chinchipe tribes in Ecuador), *ingá chata, ingá péna* (throughout Amazonian Brazil), *pacae* (in Peru and Bolivia), *ingá yowra* (among the Wayapi tribe in French Guiana), and *pois sucré poilu* (in French Guiana).

So there are at least twenty-one names for *Inga macrophylla,* which is only one of three hundred species in its genus. The total number of names for all the species of *Inga* would be roughly equivalent to the number of words in the working vocabulary of most Portuguese or English speakers. Yet these names are hardly the full lexicon for this genus. We don't know, for example, what the native people of Valparaiso named *I. macrophylla,* because they left no written or oral records.

Pennington described 300 species of *Inga,* 45 to 50 of which were of questionable taxonomy. Pennington's monograph was a Promethean work, 854 pages long, for which he spent five years examining approximately 10,000 specimens of *Inga* (including six collected by me) in museums and herbaria the world over and two years on expeditions collecting his own specimens. Before Pennington's monograph, the last comprehensive analysis of *Inga* was made by the legendary British botanist George Bentham, who published a treatise on it in 1875. Bentham described 140 species, less than half the number in Pennington's monograph. That's a net increase of about 160 species of *Inga* discovered in a century and a quarter, equal to about one and a third new species per year. New specimens of *Inga* are still being collected, probably at a pace greater than ever before in history. Should some scholar a century from now return to this genus, would she find that some of Pennington's species have become extinct? Certainly. They will have been immolated for the sake of highways, cattle ranches, onions, and commerce. Will as many more new species have entered our realm of comprehension, our lexicon? Probably not, for the simple reason that deforestation will outpace discovery. Inevitably, species of *Inga* will disappear before they are ever described by science. They will go extinct anonymously before humans ever learn of their existence, let alone of their beauty and virtues.

Inga is just one of thousands of genera of trees in the New World tropics, albeit a diverse one. And although trees may be the most conspicuous life forms in this forest and the creators of its structural complexity, most of the species here are terrestrial arthropods—insects, spiders, and their kin—or the barely described microbes such as yeasts. There are probably several thousand times as many species of the tiny beings, the invisible organisms, as there are of trees. I cannot fathom their diversity any more than I can count the stars. And, they, like the plants, are vanishing anonymously. One may argue that in this place of heroes, where children die because their parents can't afford a cheap antibiotic, who cares if a million species of beetles or yeasts go extinct, even if each, on close examination, is as radiant as a crown jewel?

I often muse on the expedition I made in 1974 over the Manaus–Porto Velho Highway. Some of the new species that we discovered then have not been collected since, and a few, I'm sure, have already gone extinct. At least we have a record of their presence on this planet. I wonder about the species—also now extinct—that we missed. We will never know how they unfold a leaf, throw a seed, arch for the light, or act in any of the other eloquent and lovely ways of being Earthlings. We shall never know Earth's language. At best, we can hope to speak a pidgin that will never mature into a complete vocabulary.

The children of northern India are taught the story of a monk—an ascetic in the sere high deserts of Tibet—who devoted his life to reciting the countless names for God. He believed that God was manifest in every part of the world, living and inanimate, even in the shadow of an insect on a blade of grass. God had as many names as there were kinds of plants and animals—and Sanskrit had a word for most of them. The monk never finished naming all the names. A journey of that magnitude could never be accomplished in a single lifetime. Like that monk, taxonomists give names to all of life's creation. Indeed, for Linnaeus, the naming of species was an act of worship. For him each name, each description, was a prayer, one of the names of God.

Some consider it an act of imperialism for a European to rename a species—in extinct Latin or Greek, no less—millennia after it was first identified by the indigenous people and entered their cultural tradition. They argue that Western nomenclature—this reverential naming of all creation—has become a rule-bound bureaucracy snarled in its own inbred lingo. The arrogance of Western nomenclature is especially obvious when taxonomists name new species after one another or after a mother, wife, or child. The critics believe that a name should celebrate a species' own virtues, not those of an alien primate.

Yet there is a sure and necessary empowerment in naming things. Words—and numbers, too—may be weightless, as insubstantial as light, yet they are terribly powerful: they can start a war, order the strip-mining of a mountain, or trigger the secretion of endorphins. And names allow us to possess our environment and manipulate it in scaffolds of thought and design.

Ultimately, the words of any language are puny tools to describe this forest. If nothing else, writing this book has taught me that lesson. I shall never be able to describe fully this painful sojourn into beauty, my tutelage in the language of this planet. I am in essence aphasic.

And now I understand why, when the Brazilian soprano Bidu Saiu was asked to describe the meaning of her music, she replied, "If I could express it in words, then I wouldn't sing it."

8

THE GIFT

O rain! With your dull two-fold sound,
The clash hard by, and the murmur all round!

    —Samuel Taylor Coleridge, "An Ode to the Rain" (1817)

LAST WEEK Tarzan and I were collecting some fungus on a fallen
log when a wisp of aroma crossed our path. I'm sure that it was no
more than a few molecules, but it was enough to get our attention.
It was cadaverine: the fascinating sweet smell of decomposition, the
bouquet that results from the digestion of protein by bacteria. It's
a molecule as ancient as dying and renewal. It portends danger,
and therefore the neurons with receptors for cadaverine are older
than thought. Every mammal has them. When an animal picks
up the scent, instinct takes over: pupils dilate, nostrils flare, breathing stops short. Soon we found the corpse: a brocket deer tucked
neatly between the buttresses of a tree. His skull and muzzle had
been crushed—the characteristic work of a jaguar—and he had
been disemboweled, the good, soft insides spilled onto the earth.
Tarzan lifted the safety on his rifle, rested it on his arm, and we continued working.

Every night the cat has returned to eat a bit more: first the brain, guts, and organs, then the haunches and face. Now only the hooves and a few long bones remain. The disappearing deer has put us all on edge. We stop to listen; we parse and interpret every sound. Before dawn there was a long scream from a nearby bush; it sounded almost human, but was just a tree frog being swallowed alive by a snake. And we watch: last night the jaguar declared her territory with rasp marks three meters up on the bole of an *acariquara* tree, and she has left soft pug marks near our camp that melt into the mud like an Escher pattern. We pause to snuffle and taste every scent that crosses our path. I've learned a taxonomy of aromas and stenches: now a molecule or two of sharp-scented urine sprayed on a tangle of lianas, the musk of a band of peccaries, the residual body odor of the howler monkeys that passed overhead a while ago, the uric acid–laced feces of a snake. The tendrils of my senses thread the dark corners of the forest and reach into the wind itself. The jaguar has presented us with a gift: the marvelous, alert perspicacity of a prey species—such an alien behavior for a top predator. To be hunted, I've learned, is to feel alive. You really can't understand this place until it turns malevolent, so that observing its details and nuances becomes a matter of survival.

Now a storm is brooding nearby. Tongues of cold, dense air lick the forest floor, following the swales and igarapés. The treetops' keening and complaints mask the individual forest sounds—the snapping of twigs and the alarms of the guaribas—that would reveal that the jaguar was near. The forest has become strangely animated; things are inexplicably set into motion. This morning, near the flyblown dead deer, we observed a tassel of radiant dead leaves suspended on a spider's thread, turning hysterically in a zephyr, momentarily looking like a bounding animal. We felt uneasy the rest of the day. Pimentel won't go near the place. Tarzan derides him, but he has started singing to himself.

※

At last, in midafternoon the storm approaches, rumbling like an oncoming train. The flat-footed trees rock back and forth on long trunks, creaking, their canopies shedding flurries of leaves. The tropeiros shriek in concert with each crash of thunder. Now the white sound of rain. There's no shelter here, no quiet place. I wonder if jaguars use the rain as an auditory blind. The storm passes over in a few minutes, roaring away, and for the rest of the afternoon the forest pings and splats, and the air, cleansed of humidity, pollen, and spores, has become invisible.

After dark, more squalls. Electricity is everywhere: purple ghosts of lightning, momentary silhouettes, unbearably close thunder. Now it is raining tree frogs, the same courting canoeiros that for the past month have been erotically croaking in the boughs above. They are leaping into a new puddle by the camp, splashing and plopping, abandoning the hard-fought safety of canopy, bromeliad, and bark for the earthly danger of the open pond, to have sex in the ozone-rich air. The canoeiros are antsy and wet, seized by the irresistible yearning for amplexus. Legions of them are pumping their gular balloons, which sound like creaking wooden hinges—faster now in the hot, dancing rain, stoking their desire. They pay no attention to me. I watch a male plaster his belly to the back of a female, holding her so tight that he dents her flanks with his thumbs. He slides his cloaca over hers, holding back but ready to release his seed. He must feel her pattering heart, her long, deep breaths, her contractions through his smooth, wet belly. Now, at once, her eggs come in long

mucousy strings: little dark planets that catch in the flooded leaf litter, chased by a cloud of a billion sperms.

Some males, in their hysteria, clasp the furry seeds of wild cash-
ews, kissing them with their cloacas, unaware that they are mating
with vegetables. But most of the unions are successful, and the next
morning the puddle is strewn with uncountable eggs, looking like
tapioca balls. Time is terribly short and the embryos must grow
rapidly; in a week each sphere will be filled with a coiled face and
tail, a beating heart. The early eggs, laid in the oxygen-rich parts of
the pond where the water is deepest, will hatch first. The polliwogs
will grow plump on the algae and fungi that coat the decaying
leaves. As my shadow crosses the pond, they will race to shelter under a leaf. As adults, these frogs will sing across the highest branches
of the forest, sixty meters overhead, and will intercept winged
moths and cicadas. But for now the polliwogs are scum-bound,
scavenging succor from the soil as fast as they can before their natal universe evaporates.

Yesterday a passing hunter brought a rumor of a floating bordello
somewhere nearby on the Juruá. The selling of comfort is one of
the more reliable professions on this frontier, profitable in good
times and bad, and wherever a bordello sets up, business booms.
The ladies hire pistoleiros to protect them, and the bordellos bring
buyers and regatões to the barracãos. They also attract preachers.
This bordello is evidently being stalked by our former neighbor
from the quay at Cruzeiro, the itinerant preacher, who follows in
the wake of sin, spinning an urticating web of guilt and conveniently proffering the balm of contrition. Last night our local workers stole away in their dugout canoes, leaving a note that they will

rendezvous with us at the Serra Divisor in a week or two. The only people left in camp are Tarzan, Dionesio, Pimentel, Zé Brejo, Arito, and the stolid Geni. As far as the work goes, we can get along without the men. It rains every day now, and the slippery wet trees are too dangerous for the mateiros to climb.

Our work is finished, but we linger here to relax, fish, and simply observe the spectacle of the rising water slowly spilling onto the land. Yesterday I watched a small tsunami, only a few centimeters high but sufficient to bear a tiny flotsam of leaves, twigs, and beetle backs, race across the forest floor. Our little camp is rapidly becoming a terra ilhada, an island of dry land surrounded by the flooded forest. Landlubbing refugees congregate on the terras ilhadas, all jostling for places in the shrinking space. The trees above our camp are filling up with scorpions and itinerant wolf spiders. With so many predators about, the canopy has become silent even after dark: to stridulate is to die. Every morning on the perimeter of our camp we find fresh footprints of peccaries, tapirs, deer, and, of course, the jaguar. The Caboclos know that these animals are easy pickings, and they regularly visit the restingas and terras ilhadas to hunt. But the terras ilhadas are dangerous places. This morning I found a plump little pico da jaca asleep under the journal in which I am writing now. I had leaned the open book against one of the stoves to dry it out; the snake must have been attracted to the heat. It was the color of earth and light, patterned like fallen leaves. Its great-grandfather might have been the one that killed Manoel near this very spot. I gently slid the snake onto the pages of the journal and carried it over to the edge of camp.

The fruiting trees of the várzea forest are ripening now, just in time to be carried away by fish or current. The green fruits of the ucuübas, the first harbingers of the flood, have already opened, revealing their bright red exocarps in which their pale seeds nestle.

Red is always an invitation to birds: now a chipping hallucination of *sete cores*—paradise tanagers—is tearing open the seeds, scattering the husks on the forest floor like pieces of bright flesh. The birds look as if they were sewn from leftover scraps of blue-green-red-black-and-yellow cloth. A *pica-pau de topete vermelho*—a crimson-crested woodpecker, its little brain protected from concussion in a fat-padded, shock-absorbing skull—batters a branch somewhere above with an audible *tok-tok-tok*. On the edge of camp, a thumb-sized *cinzento*—a yellow-crowned tyrannulet—the color of dirty sunlight, alights on the twig of a lowland coca shrub. A living thorn, it briefly stretches its wings, spreads its tail, darts into a patch of light-filled air. The brown urubús soar closer to the sun, over the edge of the forest and the riverside savanna. From the walloping midday thermals comes the clarion call, in doublets, of a *caracará*.

The Rio Azul has swelled voluptuously; its water, now a flocculent green, abounds in sediments—a photosynthetic factory. Tree boughs that once crossed open space now dip in the water. The sandbars that during the summer were the nesting places of seabirds—ospreys, black-necked stilts, terns, and skimmers—have been worn away and carried downstream, and the mud burrows excavated in the riverbank by the summer swallows have been taken over by armored catfish. In their respective seasons, both the birds and the fish excavate the holes. The birds use their beaks as shovels; the fish rub away the sides of the holes with spiny hemispheres—looking a bit like balls of Velcro—on the top of their operculum.

In a light rain, Arito and I pole *Fe em Deus* under the shade of a copse of açaí palms to go fishing. The river is angered by last night's storm, and we struggle to fix the pole in the mud and press it into the prow of the canoe as a lever. Sometimes we walk the canoe's entire length, pole in hand, pushing against the current with our bare feet. The rainwater is cold, the river warm. We drop our hand lines in at a half-meander where the current, suddenly stilled, drops its load of silt, turning the water black, as the sun sets behind a grove of spindly açaí.

The açaí may be the most graceful palm in Amazonia. Its trunk, no bigger around than a bamboo pole, grows ten meters high, supporting a drooping crown of leaves as pale green as the first hints of spring. Its leaves, feather-light, shuffle in even a modest breeze. Açaí groves are excellent for fishing; their shady understory has few snags, and their fruits, which the fish love, drop year-round. In the late afternoon sun, the reflection of their vertical gray stems is flexed by every eddy and insect, every lingering raindrop. A water strider traces its path across the reflection, across the airy spaces of the forest.

The açaí trunks are plastered with papery sheets of mud and dried algae left over from last winter's flood, and a massive spider web—a dirty shawl of silk two meters wide—is woven among several of them. No individual spider could spin such a grand tapestry; it is the colonial home of hundreds of individuals of the same species, each contributing a few strands to the collective. There is a certain economy—durability, too—to a web of this size, but ultimately each spider is on its own, hunting independently, competing to snatch the insects that blunder into the web before its neighbor does. When I stroke the web, a dozen spiderlings race to my finger. Then, realizing how out of scale I am, they retreat. Now the returning waters lap the lower reaches of the web. Tonight the spiders will move higher, crowding the tree's highest ramparts. Eventually they will disperse into the canopy. The abandoned web will be submerged, but it will continue to kill: I have observed minnows snared in these webs. In the spring, when the waters retreat, they mummify, their skeletons and flecks of scaly skin stuck in the strands of silk like silver tinsel.

After dark, Tarzan and I decide to haul one of the small dugouts over a boggy trail to an oxbow lake about a kilometer from camp and try our luck with the cast net. The night is dead still, charged with vapor, hot and juicy. An undulating layer of mist, the thickness of a banana leaf, is suspended over the trail about eye level. In the beam of my headlamp I can discern every particle of fog, herded by the heat of our bodies into little eddies that taper into the forest. The column of light excises details that are masked in the fractal daylight clutter. It draws sharp edges, tracks the route of a vine, delineates the architecture of a tree, suspends a leaf in the darkness. A spider is startled, a grasshopper pinned to its thorn. As the beam crosses a rotten log, it illuminates the crazy-legged vinegarroon who waits in the hollow center, each of its eight eyes a grain of ruby sand. A tier of brown fungus wafts a cloud of spores to the breathless air, where it lingers like yellow smoke. Hunched high in a cecropia is a white mound the texture of *papier mâché:* a three-toed sloth. It turns to face me, then bradyates to another branch, every movement viscous and deliberate. A sloth in the treetops, looking like the curled silver leaves of its preferred food species, the cecropia, is easy to miss; even its fur, colonized by algae, is tinted slightly green. It is cryptic, certainly, but I am convinced that the sloth, hunched and still on a bare branch, mimics the globular gray nest of paper wasps: swirling eddies frozen in pulp. A harpy eagle, I imagine, would look twice before impaling this lump with its talons.

On the trail ahead I catch the bright silver-yellow eyes of a spiny tree rat. It doesn't move, doesn't blink.

"*Ele está morto,*" says Tarzan. "*Cuidado.*"

Be careful: this rat, still warm, has no doubt been struck by a snake and left to die. Somewhere nearby is a pit viper.

Tarzan and I pause and observe every brown leaf and fallen twig with exquisite care. Then I see it: the red eye of a surucucú, about two meters long, at the side of the trail, its diamond back camouflaged in the clutter of the forest floor. The viper notices me, too, stops, coils, tastes the air with its tongue, dumbly strikes at the beam of my light, and recoils, rattling its white tail in the dry leaves.

It fanged the light not because it was luminous but because it was warm. The surucucú, like all pit vipers, has heat-detecting sensors embedded in crypts around its lips. A useful tool, certainly, for a predator that specializes on warm-blooded animals, but only after dark. During the heat of the day the body of a bird or mammal is about the same temperature as the air and is therefore undetectable. But after sunset, warm-blooded animals stand out to the snake's sensors against the cooling air and vegetation. Moreover, their body heat is momentarily imparted to everything they touch; the passage of every mammal and bird leaves an infrared trail on the forest floor.

The snakes know how to follow this radiant spoor, invisibly tracking their prey in the cover of night. They strike once and withdraw, letting their venom do its work. The venom does not have to be fast-acting; after all, the snake can locate the rat's warm corpse at least until it cools to the temperature of the night air. Why risk being bitten by the long incisors of a flailing rodent? Pit-viper venom is a medley of huge folded protein molecules, each designed to tear a blood cell asunder, to short-circuit a neuron, or to lyse and predigest tough mesentery. I wonder: how did these molecules evolve through a thousand small experiments, creating bonds that fold protein onto protein, adapting to the physiology of their prey, visible only by an infrared signature?

We set down the canoe and sit on it, filtering the beams of our flashlights with our shirts so as not to distract the snake as we watch it track down the rat by touching its face to the trail, pressing its crypts to the latent heat. Locating the rat, the snake butts it to make sure it is dead. And then, methodically, the surucucú unhinges its jaw and walks its fangs over the corpse, first one side, then the other, tugging at the brown fur and pulling the rat's broad shoulders into its mouth. How comforting it must be for the cold-blooded reptile to swallow the warm, radiating mammal. The snake becomes transiently homeothermic, its heat-sensing crypts momentarily and deliriously useless as the rat bleeds its posthumous warmth into its belly.

The lake is fringed by flowering bananeiras whose orange flowers capture the rainwater in pastel chalices that dance with mosquito larvae; a few brave cecropias; and tall *aninga do igapó*, arums with spade-shaped leaves a meter tall and waxy smooth, vanilla-scented flowers as pale as moonlight. These plants, like the capim, form a floating community; as we chop a path through the aningas, the vegetation undulates like a quaking bog. It is easy going; the stems are full of air.

Giant water lilies are tucked into the quiet corners of the lake. The tops of the leaves—some grow as much as two meters across—are smooth and slightly quilted, but the undersides are spiny to foil the leaf-eating predators below. The lake surface is almost completely paved with abutting lily pads vying for the light, in effect creating an aquatic understory of shade and decomposition. The water lily flowers, opening now, look like fallen pieces of moon-light. They will survive only two nights. On the first night their white petals are crenulated like Chinese rice paper. Their interiors are palpably warm and have a strong scent, like cheap perfume, that certain species of brown scarab beetles find irresistible. Moreover, the pistil is edible, studded with fatty, sweet tissues known as Beltian bodies, which have just the nutrients the beetles need. What a seduction this flower is: heat-generating, redolent, food-proffering. At dawn the flowers close, but the beetles stay, locked in soft petal and stamen, avoiding the dangers of the eye-hungry day. The flowers may be the scarabs' dalliance, but it is in the lily's best interests to evict them. On the second night the flowers open anew, but now they are undesirable: wine-purple, cold, unscented, and stripped of their Beltian bodies. The scarabs move on to new white-flowered delights, and they don't look back. Dusted with pollen they are the lily's servants now, flying penises bearing the genetic code of their seducers.

We paddle the canoe in a light rain though a copse of mungubas. The raindrops skitter over the oily surface tension of the lake, opalescent in the beam of my light. This is *água estapada*, where the water is still and its sediments decanted. These areas have a mild but unpleasant odor of sulfur, and even on a sunny day, only the top few centimeters contain any dissolved oxygen. The fish that live in água estapada must stay close to the surface, which makes them vulnerable to aerial predators such as terns and fish-eating bats. The fish with the real advantage here are the gulpers, like the pirarucú and electric eels, which swallow mouthfuls of air and absorb the oxygen through their vascularized cheeks.

I paddle in the stern, slicing the water sharply with the oar so as to make no sound, pulling hard, then using the blade as a rudder before lifting it out again. Even so, three horned screamers luff away, sounding like hysterical kookaburras. Electric eels are below: when my oar brushes one, I receive a slight tingle. A fish-eating bat, flying low, surveys the skin of the water. It uses sonar to detect the evanescent dimples made by fish coming to the surface to gulp air. A hundred thousand mayflies, as pale as starlight, are emerging from the lake, acquiring wings and leaving their larval skins floating on the water like rice husks. Wolf spiders, their hairy feet capturing a layer of air, run over the water's surface, trying to intercept the mayflies before they rise. The mayflies that survive will consort tonight on floppy wings. Just before dawn the females will sacrifice themselves; their bodies will drop back into the lake and, in a *petit mort*, they will release their eggs. Most of the males will drown, too, and their bodies will nourish their offspring.

Placing the oar handle to his ear and using its blade as a receiver, Tarzan listens for the sounds made by the fish below: the surubim makes long grunts; the pescada, short ones; the *peixe catinga* creaks like an old hinge. Now standing on the prow of the canoe, Tarzan folds the cast net over his right arm, gripping it in his hands and clenching it in his teeth. He waits: for a telltale bubble, a gulp of air, a ripple. Then all at once he flings open his arms, casting the net so that centrifugal force expands it full and round, all the while shifting his hips and legs to counterbalance its shifting movement. The canoe holds rock steady. Tarzan tries his luck five times and catches six, two, forty-seven, ten, and twenty-one fish, respectively. They are mostly pescada, peixe catinga, *curimatã*, pacu, sardinha, and piranha cajú, but also an aruanã (a beetle larva snagged in its toothed lips), and an eighty-centimeter-long surubim that vomits half a fish head.

On the next throw the net lands on top of a two-meter-long pirarucú, which surges ahead like a torpedo, dragging the canoe on a Nantucket sleigh ride into a submerged shrub. All this rigmarole terrifies the lake. Crazed silver fish scatter over the water, colliding with the boat, slapping us. The lake is raining fish into the air.

On the way back to camp we rejoin the dismal trail among the dark trees near the dead deer. Many of the fallen leaves are covered with bioluminescent white fungus, making the forest floor look paint-spattered. At first I think that the pale patches are moonbeams percolating through the canopy and dappling the earth, but this meager moon casts no light, and the canopy casts no shadow. The nocturnal din pauses before dawn; the night's screaming, eating, and lovemaking are finished. The forest is nearly silent, transpiring, dripping, pinging. We know that the jaguar is close. Perhaps she is watching us, listening, sniffing the air, tasting our strange molecules.

9

RIVER OF TERNS

"[These] provinces [are] so large yet with so few people, so distant from one another, with no police, nor reason, nor government, without principal chiefs nor obedience to anybody."

—Friar Laureano de la Cruz, 1648

ABOVE THE DESK where I am writing is a framed map of South America, a page from the Phillips *Atlas,* printed in London in 1855. The map is curled and oxidized. Political boundaries are hand-tinted purple and green. They are faded now, but no matter—the borders have changed, too. I discovered the map on Portobello Road in London, in one of those shops that butchers antique books and sells history by the page. The Phillips *Atlas* was state of the art in its time, yet stamped across western Brazilian Amazonia, across the headwaters of the Rios Madeira, Purus, Coari, Tefé, Juruá, and Javarí, is the seductive word UNEXPLORED. The rivers are sketched in, but their courses are clearly fanciful: all run straight and clean northeast from Peru, with only a few meanders here and there. None have tributaries, save for an unnamed trickle entering the western Javarí. The Serra Divisor looks like a couple of ant mounds, folding on themselves and creating a secret valley.

Perhaps, I thought, the map was just behind its time. But after spending a few absorbing days in the Map Room of the Royal Geographical Society, I learned that it was quite progressive; other maps of its vintage show no trace of the Juruá at all. Here's a chronology of the maps that I found:

1810: An unnamed map. The Rios Yutay (Jutaí) and Yurua (Juruá) drain a vast, epicontinental lake of no name. To anybody who had traveled across the mouth of these great tributaries, it would be logical to assume the existence of such a lake, but it is wholly inaccurate.

1821: Now the imaginary lake has a name, Lago Roguáguado, and is shown as the source of both the Rio Juruá and Rio Purus; the Rio Madeira originates in another fanciful lake, Lago Cayababas.

1873: Some unnamed tributaries of the upper Juruá have been sketched in, but they are inaccurate. The Serra Divisor ranges north to south and beyond the headwaters of the Purus, eastward. The Serra at the upper Rio Moa is named the Serra do Concháguay.

1892: A regression: the upper Juruá is still sketched inaccurately; there is no Serra Divisor and no place name except the Rios Mú and Gregorio.

1903: The Rio Moa is named the Rio Huapistica. The Serra Divisor looks like a series of symmetrical blobs; there are no details or settlements.

1904: The Rio Moa, without a single meander, extends all the way to the Serra Divisor (still unnamed). The Rio Azul is named the Rio Breguea.

1906: A Spanish map; the rubber boom has arrived and the Rio Juruá is studded with seringais. Cruzeiro do Sul is not shown, but upstream from its site are seringais Olivença, Matto Grosso, Carlota, Rusas, Santo Antonio, Santa Cruz, Simpatia, Puerto Camilo, Buena Hora, Cabo Esperanza, and Natal.

Although the Spanish had been systematically exploring the eastern escarpment of the Andes and the Ucayalí River basin since the middle of the sixteenth century, the other rivers of western Amazonia were among the last places on Earth to be mapped, later even than some parts of Antarctica. The valley of the upper Juruá remained *terra incognita* to both the Spanish and the Portuguese for three hundred years after the Conquest. Any Spanish explorer who entered the river from its headwaters west of the Serra Divisor was in danger of being swept through the mountains all the way to the Atlantic Ocean. No one dared to try. And the Portuguese, based three thousand kilometers downstream in Pará, considered the valley of the Juruá an indomitable hinterland, remote from the marketplace, beyond the range of their maps.

Unexplored, perhaps, but hardly beyond the influence of Europeans. Disease was the harbinger of their arrival and of holocaust. By the time the Spanish defeated the Incas outside of Quito in 1533, the native armies had already been decimated by smallpox, carried south in the bodies of sprinting messengers from Panama. Smallpox and other fast-spreading Eurasian diseases—measles, diphtheria, colds, and flu—must have arrived in the upper Juruá soon after, carried invisibly on the wind or spread irresistibly through commerce and love. The people of this river had no idea that the "New World" had been discovered, but thousands must have died of those phantom scourges in this very valley. We'll never know how many, of course. How does one measure the disintegration of peoples who left no writing or enduring monuments, destroyed by an invisible alien force that seemed to appear from nowhere and moved on like an ether?

The first documented case of smallpox in lowland Amazonia occurred in Belém in 1616, only five years after the city was founded. By 1648 the epidemic had traveled two thousand kilometers west, as far as an unnamed Omágua village on the upper Solimões close to the mouth of the Juruá, where Friar Leandro de la Cruz happened to be proselytizing. Friar Leandro reported that within a few days after the first Omágua showed symptoms, a third of the village had died of the virus.

After the epidemic of 1616, the diligent Portuguese bureaucrats recorded all subsequent episodes of smallpox that swept through Amazonia with deadly frequency: in 1660, 1669, 1680, 1749, 1756, and 1762. The epidemic of 1660 alone was documented to have killed 44,000; that of 1669 killed 20,000. Most of the dead were Native American slaves who had been abducted from settlements along the littoral várzea of the Solimões and its tributaries and taken to sugar and cacao plantations near Belém and São Luís. The slaves were valuable property, and their deaths were considered capital losses that warranted careful bookkeeping.

The Amazon had proved to be a disappointment to the Portuguese, not the font of gold and other resources they had expected. Except for a few *drogas do sertão*—the balms, elixirs, seeds, oils, and teas extracted from the forest plants—humans were about the only valuable commodity that the colonialists could extract from their vast and strange empire. Slaving is a story as old as dominion, and it's not hard to reconstruct what must have occurred after the raiding parties, known as entradas, passed through the tribal lands. The raiders received the highest prices for the healthiest, most vigorous, and fertile captives. The villages of native people were left with only the elderly and very young—those who were least capable of making a living. Cultural, economic, and linguistic traditions were torn asunder.

Along with the familiar contagions, the slavers brought an episodic new disease—yellow fever—that required vectors and reservoirs and therefore took a while to get established. Yellow fever was slower to spread but much more persistent than smallpox and measles had been a century before. It arrived in the New World in 1647 aboard a ship carrying African slaves to Bridgetown, Barbados. In Africa the virus was vectored by an urban mosquito, *Aedes aegypti* (which probably also got to Barbados in 1647), but it quickly established an indigenous New World sylvan cycle, spread by native mosquitoes and reservoired in wild forest monkeys. The disease therefore developed the troubling ability to appear unexpectedly in remote forest villages far beyond the range of direct contact with foreigners.

৯৮

Many Native Americans abandoned their traditional lands in the fertile but trafficked floodplain and fled deep into the interior to eke out a living in the terra firme forest, which nobody else wanted. Some of those tribes remain there today, hunter-gatherers untouched by Western modernity. In 1691 the German Jesuit Samuel Fritz wrote of the once mighty Omágua:

> In former days . . . [they] had been very war-like and masters of almost the whole River of the Amazons . . . but now they are much intimidated and wasted by the wars and enslavements that they have suffered and suffer from their neighbors in Pará. Their villages and homesteads [were formerly] a league and more in extent; but since they saw themselves persecuted many of them have withdrawn to other lands and rivers, so as to be somewhat more secure.

The remnant tribes who remained along the river courses fell under the spell of the West. Slowly, from both the Spanish highlands and the Portuguese lowlands, missionaries—Franciscan, Carmelite, Mercedarian, and Jesuit—cobbled together villages, known as *aldeias,* for the culturally orphaned survivors. Genocide has always been the apostle's ally: the pap and comfort of doctrinaire religion become irresistible in strange and uncertain times. In 1747 la Condamine described an aldeia in Spanish Amazonia: "Here are collected Americans of diverse nations, each of which has a language peculiarly its own, as is common over the whole continent. It sometimes happens that a language is known to no more than two or three families, the wretched vestige of the tribe destroyed."

To fill the linguistic void that resulted from the depopulation of Amazonia, the Jesuits invented *lingua geral,* an amalgam of shards of the myriad local tongues, annealed by Tupi words from southern Brazil. It was a necessary crutch for the mutually uncomprehending refugees in the aldeias. But lingua geral destroyed the survivors' ability to understand their land and its diversity. Entire vocabularies disappeared. Nobody knows, for example, the origin of the word "Juruá." Like the names of most everything here, its meaning has gotten lost in the forgotten etymologies of vanished cultures; it has become a word without a context. Constant Tastevin, a French missionary who worked in western Amazonia in the early twentieth century, speculated that the river's name was derived either from *ayuru*—a generalized Tupi-Guarani name for parrot—or *yerewa*—a name for what Tastevin called a "black gull," probably referring to the soot-colored, white-headed noddy terns that breed in multitudes every summer on the Juruá's tan beaches.

For nearly two hundred years, for all practical purposes, the Amazon was under ecclesiastical control. The Jesuits were by far the wealthiest and most influential of the missionary groups. To their credit, they opposed slavery and strove to protect their disoriented flocks from the entradas. But the Jesuits were also businessmen. Their order had no compunctions about earning a profit by enthusiastically enlisting Native Americans as cheap labor for their own enterprises. One of the most profitable of these industries was the extraction of drogas do sertão. In 1639 Cristobal de Acuña, canoeing past the mouth of the Juruá, observed that the area was rich in these products. The Juruá, he wrote, had the best sarsaparilla; healing gums and resins in great abundance; and honey of wild bees at every step, so abundant that there is scarcely a place where it is not found . . . The woods of this river are innumerable . . . There are cedars, cotton trees, iron wood trees, and many others . . . Here is excellent pitch and tar; here is oil, vegetable as well as from fish . . . there are also many others which would not fail to enrich the royal crown.

Certainly, the Native Amazonians had traditionally made a living by collecting products from the forest and river, but these items were only a small part of a resilient economy that also included várzea and terra firme farming and, of course, fishing. When the Portuguese began to export the drogas do sertão, they actively discouraged the other forms of subsistence. The Amazonians became thralls of the Western economy, dependent on imported goods rather than on their local subsistence crops, a pattern that persists today. They forgot how to use the myriad parts of the forest and river and lost the words for them, too. Instead, they extracted only those products that generated revenue for their masters.

The Amazonians traded drogas do sertão for the tools and trinkets of industrializing Europe: beads, metal knives, cooking pots, mirrors, distilled spirits. Above all, they craved iron ax heads. The introduction of this one simple item transformed Amazonia more than any innovation until the Transamazonica came along three hundred years later. Clearing a swidden plot with stone tools was nearly impossible. But a single iron ax head—an heirloom that could be passed on for generations—could bring a forest down in short order—and do so again and again.

Many of the Jesuits had been born to Iberian ranching families, and they brought those traditions to the Amazon. The combination of cheap native labor and iron axes enabled the Jesuits to clear ranches that reached to the horizon. By the middle 1700s, the Jesuit missions on Marajó Island at the mouth of the Amazon maintained more than half a million head of cattle. They were more prosperous than even the colonial government.

In 1750, under the terms of the Treaty of Madrid, Portugal was awarded formal control of most of Amazonia. Suddenly the sprawling land, which had been under the de facto control of missionaries, became legitimately Portuguese. In 1767 the Marquis de Pombal, governor of Portugal, stripped the Jesuits of their administrative authority in the aldeias and appointed his brother, Xavier de Mendonça Furtado, as governor-general of Pará and Maranhão. To substantiate Portugal's claim to Amazonia, Furtado decreed all Native Americans to be Brazilian citizens. But the shortage of labor continued to be a limiting factor to Amazonian colonial development, and Furtado's declaration was no more than bureaucratic sleight-of-hand, substituting one system of indenture for another. In fact, Furtado hadn't a clue how many people lived in Amazonia, and most native Amazonians had never heard of Portugal. He might as well have declared the trees citizens. Regardless, inducting Native Amazonians into slavery was suddenly illegal. For sure, entradas continued throughout Amazonia; the colonial authorities would never stand in the way of good business. But now slaving was off the record, for documenting it on paper had become a potential indictment.

The new governor established a system of patronage under which his cronies—known as directors—replaced the missionaries in the aldeias. The directors were afforded virtual dictatorial powers over the Native Americans in their jurisdictions and could conscript them into labor corvées for ranching and the extraction of drogas do sertão. In exchange, the directors received a percentage of the earnings. Brazil seceded from Portugal in 1822, but the system of directorships persisted. Indeed, the directors became the model for the barons and patrões of the rubber era. In 1854 William Herndon, an American naval officer who journeyed to the Brazilian Amazon scouting routes for commerce, wrote of the directorates:

> All the christianized Indians of the province of Pará . . . are registered and compelled to serve the State, either as soldiers of the Guarda Policial or as a member of "Bodies of Laborers" (*Corpos de Trabalhadores*) distributed among the different divisions (*comarcas*) of the province . . . It is from these *bodies* that the trader, the traveller, or the collector of the fruits of the country, is furnished with laborers; but . . . little care is taken by the government officials in their registry or proper government, and a majority of them . . . have become, in fact, the slaves of individuals.

Marriages between Europeans, their slaves, and the locals in the aldeias gave rise to a new type of outback Amazonian—the Caboclo—genetically a hybrid but culturally a Native American. The Caboclos had jeito, but none of them wanted or needed to live under the control of the directors. They dispersed along the river margins, living in small homesteads, fishing, hunting, and growing manioc, beans, and corn. Of course, the Caboclos' independent ways were of no use to the directors, who demanded export commodities, in particular drogas do sertão. Once again a chronic labor shortage hobbled the Amazonian economy. As William Edwards wrote in 1846: "In the vicinity of Santarem, the scarcity of laborers is most severely felt; slaves being few, and Indians difficult to catch . . . Desertion is so common, and so annoying, that it receives no mercy from the authorities."

The population of Belém slowly declined: in 1819 it was 24,000; by 1848 it was only 15,000.

※

The directorships, like the aldeias before them, had become places to die. In 1854 the English naturalist Henry Walter Bates described the pseudo-enslavement of the Native Americans in Ega (now Tefé), on the Solimões east of the Juruá:

> I saw here individuals of at least sixteen different tribes; most of whom had been bought, when children, from the native chiefs . . . They are the captives made during the merciless raids of one section of the Miránha tribe on the territories of another, and sold to the Ega traders. The villages of the attacked hordes are surprised, and the men and women killed or driven into the thickets without having time to save their children . . . This species of slave dealing, although forbidden by the laws of Brazil, is winked at by the authorities, because without it there would be no means of obtaining servants.

Bates also witnessed the effects of disease on the Native Amazonians:

> Great mortality takes place amongst the poor native children on their arrival at Ega. It is a singular circumstance, that the Indians residing on the Japurá and the other tributaries always fall ill on descending to the Solimões. The finest tribes . . . are now . . . nearly extinct, a few families only remaining on the banks of the retired creeks. The principal cause of their decay in numbers seems to be a disease which always appears amongst them when a village is visited by people from the civilized settlements—a slow fever, accompanied by the symptoms of a common cold, "defluxo," as the Brazilians term it, ending probably in consumption. The disorder has been known to break out when the visitors were entirely free from it; the simple contact of civilized men, in some mysterious way, being sufficient to create it . . . The first question the poor patient Indians now put to an advancing canoe is, "Do you bring defluxo?"

Bates became particularly attached to Oria, a little M/ ánha girl who was afflicted by the *defluxo.*

> We took the greatest care of our little patient; had the best nurses in
> the town, fomented her daily, gave her quinine and the most nour-
> ishing food; but it was all of no avail; she sank rapidly; her liver was
> enormously swollen, and almost as hard to the touch as stone . . .
> There was something uncommonly pleasing in her ways . . . she was
> always smiling and full of talk . . . The last week or two she could not
> rise from the bed we had made for her in a dry corner of the room . . .
> It was inexpressibly touching to hear her, as she lay, repeating by the
> hour the verses which she had been taught to recite with her compan-
> ions in her native village; a few sentences repeated over and over again
> with a rhythmic accent, and relating to objects and incidents connected
> with the wild life of her tribe. We had her baptized before she died,
> and when this letter event happened, in opposition to the wishes of
> the big people of Ega, I insisted on burying her with the same honors
> as a child of the whites.

10

A LAND OF GHOSTS

It is a strange thing that peoples should have so completely
vanished from the earth, that even the memory of their
name is lost; their languages are forgotten and their glory
vanished like a sound without an echo; but I doubt that
there is any which has not left some tomb as a memorial
of its passage.

—Alexis de Tocqueville, *Democracy in America* (1831)

THE MARVELOUS SKY is low and soft; sometimes a cloud breaks
loose and snags on the boughs of an emergent tree. Heading to the
Serra Divisor, we are descending the Rio Azul to where it meets the
Moa; then we will turn left and force our way upstream to the Serra
Divisor. The river has become angry, brown, and noisy, flush with
the septic things and detritus that accumulate on its bank during
the dry season: feces, decomposing remains of animals, corpses of
trees that toppled from its banks, taking along tangles of vines and
whole islands of capim. All these things have so far to go. Heading
downstream, *Fe em Deus* is nearly out of control, pushed by the big,
brown muscles of water. In order to steer, Pimentel has to keep the

boat moving a little faster than the current by throttling the diesel and lifting the long shaft of the helice over the snags. Sometimes a log turns the boat, and we career into the flooded forest, bashing into the trees. This is a dangerous business, and we have to be vigilant for wasps' nests, for low sticks that can put out an eye, for an overhanging limb that can decapitate.

Now we have entered the Rio Moa and are headed upstream. For three days we pass through a terrain with few humans, past furtive homesteads isolated by long stretches of forest. We haven't seen a fresh face for over a month, and are glad to be among our own kind once again. Archaeologists say that humans have been living in Amazonia for at least 11,000 years—not long in the history of our species, but sufficient for at least 550 generations of people (assuming one generation equals 20 years) to live and die along the course of this river. The people here remember the names of only a few of their ancestors at best. History is shallow in this place where the earliest people had no written language. They left impermanent structures of wood, vine, thatch, and mud, not the enduring stone monuments of their relatives on the Andean escarpment or in Middle America. The rains and floods have long ago erased their legacy.

The modern-day Native Americans in this region are descended from the relict populations that survived the aldeias, disease, slaving, the directorates, and conscription into rubber-tapping gangs. We shall never know how many there were before European contact, but probably hundreds of thousands colonized the edges of these western rivers. In 1905 the Brazilian Commission for Reconnaissance of the Juruá (charged with mapping the rubber estates of western Acre) counted ninety-eight named tribes that are known to have lived at one time or another in the Juruá Valley since 1700.

Today there are about thirty-four tribes (depending on how one delineates the threshold of becoming a Caboclo), totaling less than twenty thousand people.

History records the names of a few of the tribes that have gone extinct along the Jurúa: the Itipuna (last reported in 1691), Guanarú (1691), Uairua (1691), Tobachana (1691), Cauni (1691), Ururu-dy (1866), Kaha-dy (1866), Katukin-aru (1898), Náua (1906). Obviously, other aboriginal cultures that were here have disappeared anonymously. We will never know their names, let alone their traditions, because by the time pen was put to paper, the Juruá was already becoming the depopulated place that I know.

Some tribes survive only in legend and popular tradition. For example, two centuries ago long stretches of the upper Rio Juruá and Rio Moa were reputed to be inhabited by a riparian tribe named the Náua, a people renowned for their violent resistance to outsiders. "Náua" is a generic Pano word for "human," and the name continues as a suffix attached to the names of several extant tribes: Kachinawa (literally vampire men), the Poyanawa (toad-men), Chipinawa (marmoset-men), and Marinawa (agouti-men). The Rio Capanáua, a tributary of the Moa just upstream from us, is named for the Náua. Centuries ago thousands of "Náua" may have been living on the margin of this stretch of river, but their relationship to extant tribes, if any, is unknown. They must have transformed the empty restingas and beaches of the Rio Moa into villages and cropland, cut trails through the terra firme, burned swiddens, followed the piracemas into the flooded forest in canoes made of *louro preto*. Perhaps, over the centuries, the Náua made Dona Cabocla's terra roxa on the Rio Azul. Their furtive descendants may have placed the crossed arrows on Tarzan's estrada. Today just about everybody knows the name. A brand of guaraná is named after them; you can buy a bottle of Náua soda in the Barracão Aurora. But there are no longer any people who call themselves Náua on the Rio Moa. The last confirmed record of one was in 1906, in an Acrean newspaper article entitled "Last Náua Woman Marries." If any children resulted from that union, they were Caboclos.

The river, I'm sure, has looked much the same for thousands of years, although its course has slowly meandered over this flat terrain like a trickle of rain on a windowpane. But the forest? It would be easy to conclude that it has always been a wilderness where humans have been rare, a pristine place of centuries-old sentinel trees. I know, however, that it is not pristine. I can only imagine how five hundred fifty generations of innovative, hungry people have altered this living tapestry. At every crumbling concavity along the upper Juruá, the Moa, and the Azul, we find clues that tell of another reality: innumerable shards of pottery (some is polychrome, with scalloped margins), bits of bone, and the tell-tale rich *terras roxas* of ancient homesteads.

No thorough archaeological exploration of the Juruá Valley has been made, so we must use ceramics from nearby tributaries as surrogates. The nearest are from the watershed of the Río Ucayalí, where some of the earliest pottery shards conform to a style known as Hupa-Iya, part of the revolutionary Barrancoid tradition that suddenly appeared in the western Amazon about twenty-two hundred years ago. Among the Hupa-Iya ceramics are bowls of delicate and sophisticated construction, inscribed with designs that depict forest animals. They were probably used to process sweet manioc flour. About three hundred years later, the Hupa-Iya tradition was replaced by pottery of the Yarinacocha style, which is blockier, irregular in shape and uneven in thickness, and inscribed with few designs. Though crude, the large, shallow pans of the Yarinacocha reveal an important new technology: the extraction of cyanic acid from bitter manioc. Except for these hints, no one knows how these artisans lived or what they understood about this river and its forest. The banks of these western rivers have become a land of ghosts, in the words of one scholar, "an empty social landscape." The solace of this lonely river, whose tranquility so absorbs and soothes me, is, I've learned, an artifact of conquest, the noiseless aftermath of holocaust.

The first descriptions of Native Americans of the Juruá were secondhand or thirdhand accounts: rumors, contrived justifications for the entradas, or excuses to "civilize" and convert the Indians to Christianity. Slavery or conscription into labor corvées, the Europeans argued, was a better fate than running wild through the woods like animals. Consider, for example, this description, written in 1778 by Friar José de Santa Theresa Ribeiro, a Carmelite missionary on the upper Solimões:

> I certify, on priestly oath and on the names of the evangelical saints . . .
> I received a report that on the Rio Juruá there is a nation of Indians
> with tails . . . To verify this extraordinary novelty, I ordered [an] Indian
> to come to me on the pretext of taking some turtles I had gathered in
> a corral, in order in this way to check the truth. And as a consequence
> I saw, without suffering to deceive anyone, that the named Indian had
> a tail of the thickness of a thumb, a half a palm wide, covered in
> leather without hairs. And, I was assured . . . that every month he cut
> the tail so that it wouldn't grow too wide, because it grew more than
> enough . . . And, above all, to demonstrate my truthfulness, I place
> my signature and seal.

Ribeiro had never journeyed beyond the mouth of the Juruá; no literate person would do so for nearly a century. However, by his time the slavers and regatãos already knew the lower reaches of the river.

Most were Caboclos, a generation or two removed from their native ancestors, and they were fluent in língua geral. They kept no written records of their adventures, because their commerce was one of barter, not money, for which bookkeeping was unnecessary. The regatãos knew better than to seek gold; instead, they traded in drogas do sertão, turtle oil, and a new product that was beginning to attract the world's attention: rubber.

In his inaugural address on April 30, 1852, Brazil's president, Tenreiro Aranha, observed that "the Juruá has been seldom spoken of or understood" and lamented the lack of maps of the valley. Three years later the government appointed João da Cunha Corrêa, a regatão from Ega, and the "bastard brother of the merchant João Augusto Corrêa" of Pará, to the newly established post of director of the Juruá Indians, and gave him six hundred reis to conduct a survey of his jurisdiction. Corrêa's journal of the survey was destroyed in a fire; the only surviving account is contained in a letter by his son Guilherme, dated August 9, 1923. "My father," Guilherme wrote, "was a friend of the Indians . . . never making hostilities against [them], whom he strove to conquer with amity and trust." During his first reconnaissance in 1853, Corrêa canoed upstream about 2,200 kilometers, as far as the Rio Juruá Mirim, just beyond the mouth of the Moa. But he found few Native Americans to rule; the valley of the Jurúa was practically abandoned. He counted only nine *malocas* (communal buildings housing two to five families), forty-five houses, and 426 people. Still, in the years ahead, as director of that sparse population, Corrêa wove the Juruá into the web of Amazonian extractive commerce. His canoes, wrote Guilherme, "traveled downstream laden with products—*cacau,* resin, copaíba, anil, vegetable and animal oils, preserves of meat, Chilean [Panama] hats, *tucumá* as rope, oars, and some rubber."

omewhere near the mouth of the Rio Moa, Corrêa encountered the tribe known as the Náua, but they wished to have no contact with their director. When he presented them with the usual gifts, they threw them into the river. But over time Corrêa won the trust of an elderly Náua woman and her two daughters, convincing them to return with him to Ega "to be baptized into the Faith." Guilherme wrote, "The youngest daughter, whom I got to know, was named Petronilla and she told us that the Náuas didn't wish harm onto the whites, but that they had been treated badly, and many years ago their grandfathers escaped the [white man's] barbarities, having fled from their ancestral land, a pretty place near the headwaters of the river."

I wonder: what was this "pretty place"? Is it the empty terrain through which we pass today?

During the early years of Corrêa's directorship, the Juruá remained a river without maps. But by then the Europeans were beginning to notice western Amazonia and to covet its rubber. In 1867 the Royal Geographical Society, in London, commissioned the explorer William Chandless to chart the Juruá using modern cartographic methods. Chandless, who had been living in Brazil for several years, was emblematic of the intrepid Victorian explorer: fluent in Portuguese, courageous, observant, meticulous, and a bit opinionated.

The expedition was provisioned by the directorship in Ega, and Corrêa himself guided it for the first 600 kilometers. Chandless was hobbled by the chronic shortage of labor in Amazonia. "I had the greatest difficulty in obtaining a crew even for one canoe," he complained. Eventually he hired six rowers and was loaned a slave, named Dominjos by Corrêa. "Mine was Hobson's choice," Chandless wrote of Dominjos, who was "a man very useful on the lower part of the river . . . but in the crisis of the journey as bad a man as I could have had with me."

Chandless's map of the long and tapering Juruá, published by the Royal Geographical Society, is exquisitely detailed. For 1,888 kilometers—a journey of five months—Chandless recorded every meander and measured the width and depth of every igarapé and tributary, noting the erosion or accretion of the banks and the color of the water. ("Black and white waters," he observed, "differ in their general aspect as much as a negro and an Indian.") Chandless measured most distances by dead reckoning—a near-impossible task on a river where the velocity of water changes from day to day. To calibrate his map, Chandless took fifty-six readings of latitude and longitude (using a chronometer) and barometrically estimated the altitude of each site above sea level.

Like Corrêa before him, Chandless found the valley of the Juruá nearly uninhabited, with vast stretches of riverbank empty of humans:

> In all, those on the Juruá can hardly amount, I think, to 80 souls. The Juruá is very tortuous, and consequently has cut off many bends, and thus made many lakes, or rather backwaters [that] . . . never fail to impress me with wonder . . . As both the . . . Indians and the traders who travel up the river are comparatively few, the Juruá is extraordinarily abundant in fish and game . . . Cacao . . . copaíba oil, and sarsaparilla are the chief natural products.

And, he added portentously, "Within the last few years . . . india-rubber has also been procured."

There seemed to be more Caboclos than natives along the Juruá. Chandless observed, "These Indians, though they still hold together, may be considered now a part of the ordinary population of the Amazons; they all understand and can speak *lingoa geral,* and are I believe all baptized; moreover, they have a considerable admixture of non-Indian blood."

The Catauixi tribe, he noted, "are industrious and skilled in the making of pottery; but they have had much more intercourse with traders, &c., and have now but few distinctive characters. They were now all up-river, working at india-rubber for a trader."

To guide him through the middle stretches of the river, Chandless hired two Araua Indians, whom he paid in advance. "I was much pleased with their courtesy and friendliness," he wrote. "Though curious to see things, they did not beg." But after only five days the Araua men turned back near the mouth of the Rio Chiruan. "Their chief reason was fear of the [Culinos] Indians above. They are considered to be treacherous and hostile . . . Consequently it is a rule of travel to sleep on sand-banks on the left side of the river, in this part—a necessity which sometimes induced us to stop earlier, sometimes to travel later than I would have wished."

In fact, only the reputation of the Culinos haunted the river; Chandless's party encountered not a single one. Indeed, for fifteen days above the Araua village, the expedition found no trace of humans. At last his party encountered a demoralized group of Conibos, whom Chandless described as "tiresome and importunate beggars." Their village, he observed, was already acculturated, "sort of a trading post for the rest of the tribe. They work more or less for traders." Chandless hired two of the Conibo men to serve as guides. "They had no scruples about food, but a great avidity for salt; so that my stock of salt fish, which we had scarcely touched, and which my men had wished more than once to throw away, was to them the greatest of treats."

Five more weeks and no sign of people, other than an itinerant band of Catuquena men, "tall, strong fellows; only apron-clad." By now Chandless's party was getting restless: they were entering the terrain of the hostile Náua, where even the regatãos and collectors of drogas do sertão feared to venture. "Twice the oars, or paddles, were thrown away at night, in the hope of thus stopping the journey or perhaps causing a return. After that, as I had no more spare oars to lose, I always at night collected all in use and put them in the stern, and slept on top of them."

The expedition encountered another bedraggled village of natives: ten or fifteen men, a few women and children.

> Of what tribe they really were, what they were doing, and how and whence they came to the place we met them at, remained a matter of conjecture. They evidently had no fear of our attacking them, nor any intention of attacking us, for they sold very willingly their bows and arrows, and came on board my canoe in such numbers that we were regularly mobbed, owing to which it was impossible to inquire or learn much.

"One or two," Chandless wrote, "had pegs under the lower lip." But even among these remote tribes, there was evidence of contact. "One seemed to me a half-breed, having a beard and a mustache, and neither Indian hair nor type of face." Using pantomime and sketching a map of the river in the sand, the members of the mysterious tribe warned Chandless that the Náua were about a week upstream—and that they would be violent.

"At last, one evening, we reached a Náua plantation," Chandless wrote. It was uninhabited, but its size "showed that they were somewhat numerous, and its cleanliness that they had been recently there." The next day the explorers encountered the Náua themselves. "They had three very long but narrow canoes, and at once ran to these, beating their breasts with their hands, and their large round black shields . . . with their spears . . . and bows and arrows . . . From their gestures, I can scarcely doubt that they came to fight, but I doubt a good deal whether at first they knew we were white people, and I hoped by showing them beads &c., to bring them to parley."

Though Chandless may have had good intentions, his men were spooked and turned trigger-happy. "But when they were eighty or ninety yards off, and still making signs of war, my men . . . would wait no longer, and despite my orders fired before an arrow had been shot at us. The first shot missed, and they still came on, but the second shot wounded one of the Indians in the arm, and then they stopped, and at a third shot retired."

The Náua retreated to the bank and hid in the forest. "One arrow only was fired at us, and that fell short." Chandless's men were now in a state of mutiny. "At the proposition of continuing our journey, there was a general outcry . . . To fight our way, or even resist continual skirmishing attacks, we were too few, being only eight in all."

The exploring party turned back just downstream from the mouth of Rio Ipixuna. Chandless never made it as far as the Rio Moa or the Serra Divisor. "But I should have gone on and seen," he lamented. "As it is, I look back on the day with shame."

After Corrêa brought the valley of the Juruá to the world's attention and Chandless accurately mapped it, everything changed. Clearly, the vestigial Native American and Caboclo populations of the region would be insufficient to meet the labor demands of rubber tapping. In the 1870s the empty river margin started to fill with thousands of new immigrants from Ceará. By 1905 the Brazilian Commission for Reconnaissance of the Juruá recorded 325 rubber tappers' barracões between the mouth of the Juruá and the Peruvian border, including the first European settlements on the Rio

Moa. Although the commission listed the tribes on the Juruá and its tributaries, it provided scant information on their lifestyles or numbers.

The first accurate ethnology of the Native Americans of the upper Rio Juruá and Rio Moa was written in 1913, four hundred years after the European conquest. The author was Friar Constant Tastevin, a French anthropologist and missionary at the Congrégation du Saint-Esprit in Tefé.

Tastevin made seven journeys up the Juruá between 1908 and 1914. He wrote lyrically of natural phenomena, and for this reason alone, his accounts are worth reading. He extolled "the vast, immaculate beaches of the Juruá . . . brilliantly pale, the clear sand of an impalpable fineness, the waves that are stained by wash from the stream or the wake of the canoes entrained and folding along the beaches, dying lapping like waves of the sea on the beach."

Tastevin described a "veritable train of forest along either bank" and "the [vast] numbers of trees toppled across the river," which created "pseudo-rapids." Thousands of *tracajá* turtles [now nearly extinct] hauled out on the snags to bask in the sun. Although the oxbow lakes were "of dead water," they were "a paradise for caimans . . . and the infinite varieties of fish." The waters of the Juruá, Tastevin observed, were "red, white, yellow or violet," depending on the origin of the nearest tributary. "The black waters," he wrote, "appear like clear and limpid glass, glazed with pale amber."

Noting "the extreme sinuosity of the Juruá's course," Tastevin described how the locals measured distances, not by kilometers but by the number of meanders and beaches: "It is eight beaches distant: three large, two small, and three middle-sized, plus two *estiroes* . . . Go to where the waters run yellow, but the river seems blue to the horizon."

On Tastevin's penultimate journey up the Juruá, in November

1913, he entered the Rio Moa and paddled upstream as far as Gibraltar. The trek took Tastevin and a companion eight days, rowing twelve hours a day, over the same route we are taking nearly eight decades later. He knew that the Juruá originated "at the foot of the nearest hills of the Peruvian Andes," on the other side of the "rocky, wooded hills of the Contamana [Serra Divisor]." The Serra, he wrote, was "covered with virgin forest which gave it an azure color and a grand aspect."

These vistas appear much the same today. But Tastevin the ethnographer and linguist (he learned the local Panoan languages in order to "teach to the Indians the beauties of Heaven") provides us with a unique window on the lifestyles of the Native Americans of the Juruá at that time and a disturbing insight on their demoralization and decline. He recorded twenty-one native tribes inhabiting the upper Juruá—more than on any other part of the Amazon. Most lived in small homesteads of a family or two, separated by long uninhabited stretches. "Malarial fever and, above all, beri-beri are severe, cruel and continual," he wrote. The natives, he found, were "on the road to extinction, due to misery, disease, and above all the gripe and catarrh, as well as voluntary sterility."

Tastevin was surprised to learn that some of the native people went to great lengths to conceal their true tribal identity—a cultural camouflage that had protected them from the directors and patrãos. He observed this denial in a settlement of natives who claimed to be Katukina. Tastevin could tell, however, that they were not speaking Katukina, but a dialect of Kachinawa.

"Are you really Katukina?" Tastevin asked Mame, their chief. "What kind of people are you?"

"We are Katukina!" Mame replied.

"Then why do you speak the language of the Kachinawas? Are you not Kachinawa?"

"'We, Kachinawa? But the Kachinawa are our enemies . . . [Tl are] cannibals, murderers, thieves, sloths. On the other hand we are brave people, we don't harm others and we are good workers: behold our fields, our maize, our manioc, our banana trees!'"

Tastevin pressed his interrogation:

> Backed into a corner, Mame admitted that he was Kachinawa . . . Why, then, do they call themselves *Katukina?* One supposes that it was to avoid persecution by the whites. When whites appeared in the country, they were accompanied by Katukina . . . who had been the friends of the *civilisés* for a long time . . . Hundreds of [Native Americans] were massacred without pity by the *civilisés,* and also by the half-civilized in Peru. To escape these killings, the *Kachinawa* proclaimed themselves *Katukina,* and no longer wanted to be known as affiliated with the *Kachinawa,* which were the principal victims of the massacres.

Years before, when I worked at Santa Luzia, our closest neighbors were a settlement of Katukina—at least, that's what they said they were—about forty kilometers farther east on the Transamazonica. Arito had befriended the tribe about a decade earlier, when he was a bulldozer driver in a road gang building the highway. That particular stretch of road was wild and dangerous then. (One night a bus driver, squatting to defecate on the side of the road, was killed and partially eaten by a jaguar, the back of his head crushed in characteristic fashion.) Until the highway cut across Acre, the Katukina lived on the margin of the Rio Jordan, but they had moved to the road because it brought commerce and other interesting things. The road also brought disease. Within two years forty-eight Katukina children had died of measles.

"There is no more game, no jabutí, no guaribas," their chief lamented to Arito (whom he called "Capitão," an honorific reserved for rubber barons a century before). "We eat jacarés now. Imagine!"

Regardless, the Katukina remained on the roadside, in a settlement where a wooden bridge crossed an igarapé. Over the years, as the highway slowly deteriorated and the bridge rotted, they were isolated once more.

But in the mid-1980s the road was still passable, and the Katukina village was an easy jaunt from Santa Luzia. The people there accepted me because I was Arito's friend, and we often visited the tribe on our days off. By then the Katukina preferred Western clothes, had portable radios, and made money selling farinha and dried fish. But most of their income came from putting on mariri ceremonies for the city folk of Cruzeiro do Sul, members of the União Vegetais, a cult made up of a mixture of Catholicism, Protestantism, African Umbanda, and a belief in "vegetal powers," who used drugs as a source of religious inspiration. The purpose of a mariri was to drink *santo daíme,* a hallucinogenic decoction made from the stem and bark of the caapi vine. Its recipe was the Katukinas' secret, and they charged a hefty fee for their ceremonies. Every mariri I observed was a business venture.

The sponsor of one of the ceremonies I attended was a wealthy businessman from Cruzeiro, who stopped at Santa Luzia and gave us a ride to the Katukina settlement.

"You're a botanist, no?" he asked me. "Well, then, of course you understand the mystical powers of the forest," he continued conspiratorially, as if he and I shared some special knowledge. "You understand that we seek primal knowledge and inspiration in the chemistry of forest plants. They are gifts of the forest spirits, put there for a purpose. Why, after taking santo daíme, I've personally had visions of ocelots and jaguars. I believe they are my totems. They help me in my business dealings.

"You see, our everyday lives are mired in the Corporeal Spirit," he declared, tapping his all-too-corporeal belly. "The forest holds

the secrets of the Ethereal Spirit, and they are revealed by the p
of the jungle. The marriage of female *café brava* and male
gives birth to inspiration."

I knew that santo daíme was a poison made up of a medley of
molecules precisely crafted to bind to neural receptors, short-circuit
them, and incapacitate any mammal dumb enough to eat the plant
it comes from. It's an effective strategy for the plant to eliminate its
browsers: a spaced-out deer or tapir has a good chance of being
eaten by a jaguar.

In one of the village houses, João Minhoca, the Katukina sha-
man, allowed me to watch him cut the sections of caapi vine and
shred the leaves of café brava, which he boiled together in a kettle.
I recognized the plants, so there was no point in trying to keep them
a secret from me. Besides, João knew I had no interest in taking the
drug. After fifteen or twenty minutes he strained the fluid through
a cloth and ladled it into tin cups. The broth had a pleasant aroma,
like cooked beet greens. João allowed me to photograph the prepa-
ration of the broth, but once the decoction was strained, photo-
graphs were forbidden.

"A flash would steal its power," he said.

The mariri began at sunset on the mud beach where several families were roasting jacaré tails and a *mutum* over clay hearths; the
Milky Way seemed to merge with the sparks from their fires. I sat
on the edge of one of the hearths and watched wordlessly. After a
while some children gathered around me; the older ones carried
younger brothers and sisters on their hips. They giggled now and
then, curious . . . slightly afraid?

João tasted the santo daíme, declared it ready, and gave each of
the members of the União a cup.

"Sip it slowly," he ordered.

At his cue, several Katukina men and women began a repetitive chant, as monotonous as a mantra. The União members moseyed to the water's edge, stripped down to their underwear, and lay on the beach, listening. After a half hour or so, the drug started to take effect. Some people started to retch, then passed out, awoke, retched, and passed out again. By midnight most of the trippers had awakened a second time, afflicted with uncontrollable liquid diarrhea. They raced to the water. One, too nauseated to walk, squatted in his poopy pants.

At that point the hallucinations set in, and for the rest of the night the participants moaned, retched, joined the mesmerizing chant, sat and stared. From my alkaloid-free perspective on the sidelines, there was not much to it: incoherent, aphasic muttering, a dying people, a dying forest lacerated by this obscene highway, a New Age businessman. These urbanized people, I thought, understood the forest even less than the colonists on the Transamazonica; they may as well have gone to the opera.

By dawn, everybody was asleep, some sprawled in their own filth.

"Did you have a vision?" I asked the businessman after he awoke.

"I don't remember for sure. A tapir?"

"And you, Senhor," I asked João. "What did you see?"

"My people no longer have visions of jaguars and ocelots when they drink santo daíme," he replied. "My people now have visions of trucks and cities."

I photographed every Katukina in the village that morning, promising to bring them prints when I returned the following year. They had seen photos of people in newspapers and wanted some of themselves. I exposed roll after roll of film, making a record of all their faces. For years afterward, I contemplated those faces and wondered whether the tribe had stayed there or had retreated into the forest or moved to the city. The photos are a last-minute portal to a remnant of pre-Columbian Amazonia. A few of the faces were already Caboclo: a young man with a scruffy beard, a boy with gray eyes, a lovely teenage mother, shy, smiling ever so slightly, nursing a runny-nosed baby.

I kept my promise and returned to the Katukina village the next year. Walking from hearth to hearth, I passed out the portraits, matching then with the faces I saw—all except the young mother and child. When I showed their photo to the other Katukina, they recoiled as if I'd handed them a ghost.

And then João explained to me what I had feared: the mother and child had died of fever. From his description, it sounded like yellow fever, for it appeared from nowhere. Cruzeiro do Sul had a hospital and a supply of vaccine, but the most recent cohort of children hadn't been immunized.

Nobody wanted to keep the picture of the girl and her child. Nobody had the nerve to behold it.

11

ANGELIM

Suddenly the sun rose—
the scent of plum blossoms
along a mountain path.

—Matsuo Basho

ON THE AFTERNOON of the third day after leaving Valparaiso, we reach República, a Nokini and Poyanawa settlement on a crest above the biggest bend of the upper Rio Moa, which commands a sweeping view of the tawny sandbars of two vast meanders. Nobody, friend or enemy, can approach this place without being spotted kilometers away.

República. No word has more cachet in western Amazonia. Founded in 1910 by Native American and Cearense rubber-tappers after the *nordestinos* expelled the Peruvians, the settlement has became a polyglot, polychrome asylum for vestigial Indians, escapees from the seringais, regatãoes, bandits, and abandoned children.

República. A name rewrought by the history of two cultures, Cearense and native. Here the various tribes of the Rio Moa took their last stand, expelling both priests and patrões, defying the entropic forces of conquest. Here they established a place of refuge, of self-rule at last, a commonwealth.

República is freedom.

From afar, not much distinguishes República from the other settlements along the Moa: there are clumps of banana trees, a generous spreading mango tree, a modest barracão, a chapel of the Cruz Vermelho, a cluster of piebald cattle on the bald land. It is home to about 150 families, including a few *brancos,* Caboclos, three teachers, and one nurse. Most of the houses are thatch-roofed, but eight have tin roofs, and the barracão is roofed with tiles.

Nailed to a wooden fence is the sign:

> NATIONAL INDIAN FOUNDATION
> (FUNAI)
> NOTICE
> TO SELL OR SERVE
> ALCOHOLIC DRINKS TO INDIANS
> IS A CRIME.
>
> Violators are subject to a penalty of jail for 6 (six) months to 2 (two) years (Section III—article 58 of law 6001/73, Statute of the Indian)

On a frontier where there are no schools, where children can't get antimalarials, where there are no police, and where refugees from the Transamazonica who have no jeito settle without plan, suddenly the government intrudes. This is Indian Country, where bureaucracy has always been the silent tool of control. From the establishment of the Republic of Brazil in 1889 to 1988, when a new constitution was adopted, Native Americans were considered wards of the state, with much the same legal status as children, deemed incapable of looking out for themselves. How could a people who named the parts of this forest, who for millennia made a living from it and who know how it all goes together, be considered wards of the same state that methodically destroyed this place?

⬙

Dona Ausira, República's Nokini matriarch, lives in a pink cottage on top of a hill at the end of a pathway of guavas, annattos, and lantanas, shaded by an old mango tree; alert gray cats and a tethered squirrel monkey climb in its branches. Its walls are made of crudely sawn lumber, the floors of split buriti stems hand-hewn with a machete. Inside one sees a red crucifix, portraits of hirsute saints, machetes, pot and pan lids, a hammock with turquoise mosquito netting, tennis shoes hanging on the wall; bundled corn cobs hang under the eaves of the sooty thatched roof.

Tonight Dona Ausira has invited us to her house. Sitting straight-backed in her rocking chair, its armrests rubbed brown and smooth from generations of use, she wears a golden plastic bow in her hair. The rest of the settlement has come here to see us, too: Dona Ausira's children, grandchildren, nieces, and nephews, as well as orphans and travelers. All gather around her, squatting on their haunches or sitting on the floor, their faces sketched by the lambent light of a lantern. They are all watching me.

"Dona Ausira," I begin, "how long have you lived here?"

"Well, nobody is actually *from* República, you see. This is a place to which we fled.

"I was born on the banks of the Río Javarí, somewhere near Jacerana, Peru. I don't know exactly when, but it was probably about 1905." The Javarí valley was a notorious place then, where Peruvian rubber barons usurped the lands of the Nokini and other tribal people for their seringals.

"We were hunted like animals," Dona Ausira continues. "The

coreiras were like a game—a sport—for the brancos. There weren't many of us to start with—only about fifty—living as nomads. We were being killed off."

And so, when Dona Ausira was a toddler, her father and mother, along with a few relatives and neighbors, fled across the border to Brazil and the valley of the Rio Moa.

"We had to trek across the Serra Divisor from the Ucayalí valley to the Moa valley. We couldn't follow the river courses because that's where the hunters were. We walked through the forest for weeks. I walked, too. My mother was too weak to carry me."

The flock first settled at Seringal Timbauba, below the Cachoeira Formosa, but soon succumbed to the debilitating routine of rubber-tapping and debt.

"My brother and sister died of malaria and measles. After a while we had to move on.

"Then we founded República, a place of our own. It saved us. This is *nossa terra* now."

During her long life at República, Dona Ausira bore twelve children; seven are living. All are Caboclos, speaking Portuguese with only a vicariant knowledge of Nokini. The oldest, Terezinha, lives in São Paulo. This year Dona Ausira adopted Erundinha, a three-and-a-half-year-old African Brazilian girl, whose mother died in childbirth.

"Why do you wear the cruz vermelho?" I ask.

"It was a gift of Hermão José himself. Although he never visited República, many of us traveled to Mancio Lima, on the Rio Juruá, to see him. That was about 1960. He had been wandering in the forest among the homesteads and villages along the estradas and animal trails," she explains.

"As was his way, he never accepted a ride in a canoe. He was never observed to eat, even by those who traveled with him. He must have been about forty years old then. He was a beautiful man with a ragged black beard, but his countenance was sad. He carried a red staff, wore a dirty white robe with a blue sash, and sandals; there were many bites on his legs. And palm fronds. Hermão José always carried palm fronds, which he cut in the forest, and he lugged a red cross with three hinges, slung in a sack on his back. It was as big as a man. When he got near a village, he unpacked the cross, unfolded it, and dragged it into the village.

"One day he packed up and left. Ever since, I've worn this cross around my neck."

"Why?"

"He seemed to *know* things. I was losing things. I lived in a place where there were no people who remembered the old names."

By the time Hermão José wandered across western Acre, Dona Ausira and her three brothers were the last speakers of their native language, as far as they knew. There may have been other Nokini somewhere—an isolated family, perhaps, at the end of an unnamed tributary—but they were out of touch.

Now only one of Dona Ausira's brothers is left, and he has few memories.

"Do you remember the word for star in Nokini?" I ask.

"*Ishtí,*" she replies without hesitation.

"The word for moon?"

"*Uhnuk.*"

"For bird?"

"*Ceu nikt.*"

"Fish?"

"*Natú.*"

Tonight may be the last time these words—which once symbolized heaven and river and forest to her people, which gave those who spoke them the perspective to contemplate themselves and their homeland—would ever be spoken. I may be their only scribe.

Dona Ausira has turned silent, her chair creaking on the floorboards. I study her wrinkled face.

"Was there a word for love?"

"There may have been, but I have forgotten."

What is a forest without names for its parts? Is it diminished? Each culture has its own taxonomy to sort diversity, to assign it order. Names are power, and they endow those who know them with understanding. But the Rio Moa is a place without names. There are no longer any native people who know their way through the várzea and terra firme and have identified its parts. Only strangers live here: furtive wanderers, refugees, monomaniacal tappers of rubber, bureaucrats who preserve a wounded wilderness that they neither understand nor like. This forest and river have no human voice to represent it, no scaffolding of words on which to hang things. There can be no psalms, no prayers here. We scientists, equally blind, are only beginning to grope our way through the undescribed. We have started to endow the forest with voices, but it's a long journey. How, therefore, can anyone achieve wisdom in this place?

And this lack of language may explain the terracide I see around me. It's hard for people to love a place that is not defined in words and thus cannot be understood. And it's easy to give away something for which there are no words, something you never knew existed.

I mourn the passing of these words on the verge of extinction. Each is an evolved being, its history etched in phonetics and rules of expression, just as the morph of a tree is patterned in the base pairs of its genome. The phonemes of these words were present when Ausira's ancestors crossed the dry Bering Strait. They were tongued and slowly transmuted by seven hundred generations of wanderers down the spines of North, Central, and South America. They embodied the memories of prairie and forest, of sea and mountain. Dona Ausira's language is a phantom, her culture a dream.

"Does it make you sad that you are the last one to speak Nokini?" I ask.

"For a while it did. Everyone who spoke it had died, and then there was nobody to talk to. I was very lonely. I was surrounded by people who were no longer pure Nokini, but who spoke Poyanawa or Portuguese. It took me a while to learn to speak again."

In a place where every mote of green life reaches toward the light, I realize that I have forgotten to ask her the word for sun. But I check myself. What would be the point of asking? Dona Ausira waits patiently for my next question.

It doesn't come for a long while.

"Do you still dream in the old tongue?" I finally ask.

The impertinence of my question is immediately clear.

"It doesn't matter, don't you see?" she chides gently, extending her hands to those gathered around her. "*These* people are my family now." The changeling girl, whose African ancestors had lost another cosmos during their own diaspora, climbs into her lap.

"This girl is my daughter now," Dona Ausira explains. "A family is not words, is not even blood. Everybody is new on the Rio Moa." She gestures to the gathered children. "A family, you see, is place. A family is República."

We can sense that the mountains are close. The Rio Moa has become shallow and swift; its water, born in artesian springs, has turned clear and cold. The river bottom has changed from pale mud to sand and now tawny gravel, which marches downstream in moving ridges like washboards on a road. Then, beyond a meander identical to a thousand others, two novel things appear: the Serra do Moa and open sky. The Serra is no more than a few green nubbins in the distance. No matter; it will grow. But the newfound horizon, the clear and distant perspective, is mesmerizing. We all pause to contemplate it. Like a deep draft of air, it seems, momentarily, to clear the head.

We have arrived in cattle country, an expressionless, captive terrain of fences and barbed wire, scruffy grasses, a few blackened stumps of trees. The cattle are few, lean, and bony—this is not a generous land. It's a story as old as the New World: the bovine monotony of Iberia transplanted to the Americas, the sepsis of people living with their domesticated animals, the ever-present stench of excrement, flies, the swagger of lonely men on horseback, the greasy satisfaction of meat. But it has never been easy to raise cattle on this frontier; men's backs have been broken here too.

Around a bend is the cattle ranch of Francisco José de Souza, better known as Bolota. Born in Ceará, Bolota was brought to Acre by his parents when he was three years old, and he settled on the Moa in 1945, when he was twenty-nine. His passage into the forest was by a common route: he first worked as a regatão on the upper Juruá for four years, then extracted drogas do sertão in the Serra Divisor for five years, and finally, having wrangled title to six hundred hectares from the territorial government, he settled near República. Living among the refugee Nokini, who were always suspicious of outsiders, Bolota refused to carry a pistol or a rifle—a conspicuous vulnerability that may have been his best defense. He went into partnership with Cacique Umberto, the Nokini leader, and became the common-law husband of one of Dona Ausira's daughters, Rosalia.

"In those days," Bolota tells me, "most of the items we wanted—gunpowder, machetes, utensils—were imported by ship, sometimes from as far away as the U.S. We learned to grow everything else, even chinchona. Malaria, you know, killed many of us at first. People shivered so much that their hammocks wobbled."

Before long Dona Rosalia had presented Bolota with a son and five daughters. The couple owned twenty-five head of cattle; Cacique Umberto owned ten more. Bolota planted a field of sugar cane, built a mill, and began to make molasses and *rapaduras,* pungent black sugar cakes so sweet that they make your teeth ache. República's farinha came to be regarded as the best in Acre; it had the fattest pigs, too. Every year at high water, Bolota and Umberto piloted a barge filled with produce—as well as live pigs and steers—downstream to market in Cruzeiro do Sul and returned laden with supplies.

By the 1980s Bolota and Umberto were prosperous. "We were both learning a new way of life," Bolota says, "and we needed each other." República had become a center of enterprise and of conviviality between whites and Native Americans. Most of República's communal structures—the three-room school, the clinic, the well—were built by the two men.

Bolota has cut stairs into the soft sandstone of the riverbank—the first folds of the Andes—leading up to his homestead. The yard is a babble of hens and their chicks, as well as five sheep and a billy goat, all guarded by an ebullient blond bitch named Xuxa. The house looks like Iberia transplanted. It has white stucco walls, blue trim, a butter churn, and a few local touches—floors made of striped *sucupira preta* and *amarelinho* wood, no furniture, just hammocks—and a gas refrigerator.

Bolota, slightly deaf, compensates by shouting everything. As he serves me a cup of yogurt, tipping back his Panama hat and tapping my elbow, he loudly complains, *"Why do you keep asking me the names for things!?*

"Ecologists like you interfere with the nation's destiny," he bellows. "What could you—a city boy—possibly know about the practical things of living in this place, understand of our struggles here? *Compreendeu?* Why, do you even know the cost of a sack of cement? Aha! I didn't think so."

Recently, 45,000 hectares surrounding República were declared to be a reserve for the Nokini, and upstream, in the Serra Divisor, 809,000 hectares have been designated a national park.

"Now the government is taking our land for reserves, making life impossible for poor pioneers like me. It doesn't matter that we raised our families here or that we buried our children here. Compreendeu? Those of us who can't prove we are natives are out of luck.

"What difference does it make whether we were born here?" he asks. "We're all Caboclos, you know. Even Cacique Umberto doesn't speak Nokini any longer. He has become a Caboclo too.

"Look, I'm no doido," Bolota continues, stabbing me with a finger. "I know more about life on this river than you ever will. And I know answers to questions that you haven't thought of yet."

I shrug. He's right, of course.

Bolota steps back and crosses his arms. "I'm book-smart, too. Why, I've educated both of my sons. One is a professor. The other is normal."

Dona María Luiza de Oliveira arrived in República in 1952 and has been the nurse there ever since.

"I have delivered 1,258 children and successfully treated thirty-nine snake bites," she says proudly.

Last week a four-year-old girl was bitten on the foot by a suru-cucú. Dona María first flushed the bite with water to keep the girl's circulation going.

"I used a tea of *cumarú cheiroso* bark as a coagulant to stop the bleeding," she explains. "Dona Ausira taught me that. She still knows a bit of Nokini medicine, but she is the last one to remember it."

As we journey closer to the Serra, the forest gives way to contiguous pastures that vault over the horizon: Eurasian grasses, Eurasian grazers, Eurasian browsers. There's a systematic biotic cleansing going on here. The settlers clearly don't like this place and are repelled by the forest. They seem to be living in exile. None of the ranches, big or small, is very prosperous. No one is doing very well. Why do they bother to grow cattle here, so far from market?

"It's because of the park," explains Arito. "Only speculators live here now. They figure that if they invest heavily in ranches, clearing the land, and raising cattle, the federal government will have to buy them out at a significant profit. It's illegal, you see, for people to live in a national park. All of these people will have to leave."

"Where will they go?" I ask.

"Well, anyplace is easier than here, I suppose. With the money they are paid, they will be able to buy a house in the city."

We've reached the Serra at last. The river bends sharply south through a shadowed canyon of mealy green sandstone cliffs into which it has cut eaves stained red with algae. It's suddenly chilly here in the shade. The valley of the Serra, its face browed with ferns and cyclanths, seems like a secret realm at the end of the earth. The mountain faces are dark gardens, mist-fed forests of tree ferns and small shrubs, a steep and tumbling canopy of bearded gray lichens, pale vines, and wet fields of grass. Curtains of roots from the trees above are draped over the cliff face. At this season the river climbs tiers of chortling rapids, where we have to pull the canoe against hard, cold water. Each rapid makes islands of brown spume, where organic matter has become trapped in soft saponin bubbles, the residue of decomposing plants. Long before we see it, we can hear the waterfall reverberating through the canyon, its white sound fading and returning as it is absorbed by the waves of the day's spent heat rolling over the river. Around the last bend we find a tawny cascade two meters high, a tangle of snags, rainbows in the new mist. We may be in Peru now; there is no way to know. Six months from now, when the river is swollen, we could easily slide over the waterfall to the other side of the Serra. But today this waterfall is the end of our journey.

The first explorers arrived in the Serra do Moa in 1936, when an expedition from Petrobrás—the Brazilian national oil company—drilled some test wells in the mountains. The drill was steam-powered, the boiler fired by native wood. Instead of oil, the drillers struck an artesian well. Today the rusted boiler and the perforation in the rock, gushing hot, sulfurous water, are still here. Soon afterward, the first Caboclos arrived at the Serra. By 1960 nearly fifty families lived in the area. They subsisted by hunting, fishing, and growing manioc on a shifting mosaic of swiddens. But they did not tap rubber. The Serra do Moa is one of the few places in western Acre with no wild rubber trees, and therefore there were no landowners, no peonage to a patrão. There was freedom here. The homesteads were only a hectare or two, enough to support a family. But a hectare in this part of the world has eight hundred to nine hundred trees, and few of the pioneers had enough money to buy a chain saw.

The patriarch of Fazenda Arizona lives just downstream from the rapids. He has built his house high on an eroding bank so that he can keep track of who is portaging up and down the river. Francisco Xavier Barbosa de Lima, known as Chiquininim, arrived in the Serra do Moa in 1972, following the path of his older sister, República's legendary nurse, Dona María. He was well educated, having spent three years in Belém and seven years in Manaus, where he started his residency in medicine but, finding he had no stomach for the profession, he dropped out. Chiquininim was a different kind of settler: visionary, urbane, highly capitalized, and willing to risk it all. Using borrowed money, he purchased about 500 hectares of forest at the base of the Serra. Hiring laborers from República and the neighboring seringais, he slashed and burned about half of his spread during the first dry season, sowing the denuded land with coffee, guaraná, and rubber plants. A year later he introduced pigs and cattle. At first the plantation produced no food. The men hunted along trails in the várzea forest during the dry season and canoed over the same paths during the wet season. Amazonian dugouts have no keels, and Chiquininim, paddling alone from the prow, had to put a stone counterweight in the stern. Soon he married a local girl, María Antonieta Herreira de Souza, and started a family. Now he has six kids, and he doesn't need the stone.

It was tough going for Chiquininim and María. One of their six children died of dysentery brought by the floodwaters, and when travelers came, they inevitably brought malaria. Armed drug smugglers crossing the border from Peru were a constant danger. Jaguars regularly raided the pigsty. To keep them at bay María maintained a fire in the clay hearth all night long. But by the mid-1980s, Fazenda Arizona had 290 cattle, 500 or 600 pigs, 20,000 coffee shrubs, 1,000 guaraná vines, and 54,000 rubber trees. Farming rubber trees is always risky in Amazonia because of pathogenic fungi, but Chiquininim—ever the visionary—figured that because there were no wild rubber trees in the Serra, his trees might be spared. Although it takes only seven years for a rubber tree to yield latex, a lot of labor-intensive weeding and trimming must be done during that time. At its peak, Fazenda Arizona employed more than one hundred men, but there was always a shortage of workers—and of money to pay them.

We camp at Fazenda Arizona, stringing our hammocks from the beams of an open-walled house surrounded by a lumpy meadow of African kikuyu grass, with lowing cows and their attendant bull. But as soon as we land, we encounter pain and responsibility: a nine-year-old girl who has been stabbed in the foot by a stingray. She is the daughter of Senhor Tanti, a burly neighbor whom I've met before. Last year we treated her for double malaria and hepatitis and took her to the hospital in Cruzeiro do Sul.

The gall in the girl's leg is twice the size of her ankle, making her appear club-footed. She can't walk on the injured leg and has to be carried by her mother and laid on the floor of our house. The wound is stretched and shiny, the thin white scar of the stingray's barb etched on the skin like an albino tattoo. The pus is completely enclosed, as if her body were trying to bury the insult.

Chiquininim sharpens his Swiss Army knife on the broken edge of a cachaça bottle. He holds the blade in a candle flame, contemplating his route through the skin into the wound.

"I don't know where the nerves and blood vessels are," I say. "Perhaps we should just let it erupt on its own."

"No," Chiquininim replies. "She may get a fever and die. All we have to do is cut gently across the face of the boil, and the pus will find its way out."

By now the sobbing girl has decided that she doesn't want to proceed with the operation. But her mother holds her down, pinning her legs to the floor of the hut. She has had to do this before. Chiquininim scores the wound, each stroke of the knife as thin as a paper cut. The slice spreads rapidly because of the pressure inside. When it fills with blood, I dab it with gauze, and Chiquininim slightly changes the trajectory of his strokes. The girl, surprised that there is no pain, has stopped struggling and watches. In fact there could be few living nerves in the dying flesh above the wound. We pause. A moth incinerates itself in the kerosene lamp, creating a smudge of smoke. The girl's brothers and sisters, unexpectedly fascinated by their sibling's misfortune, are absorbed.

Another stroke—and suddenly the pustule erupts. It makes an audible pop, flooding the lesion and spilling over our hands. Blood and pus—smooth and creamy as mamey fruit—spatters us all. The smell is oddly earthy. Now the girl screams again, as her tissues and bones, released from the pressure, shift to their correct positions. Lest she contract tetanus, we place a cotton wick in the wound to aerate it. The girl sleeps.

At dawn we wake to more pain: the rantings of Sebastão the madman in the death-still air. He lives next door, locked naked in a cage of split logs behind his family's kitchen. We offer our help to Sebastão's parents, Senhor Francisco and Dona Justina, as we sip coffee in their kitchen. As if he were double-jointed, Sebastão twists his face into a gap under the eaves to watch us. His crooked fingers are as white as porcelain, his long amber fingernails poke between the slats. I peer into his eyes, but he looks away. Sebastão has gnawed the windowsill of his prison until it is brittle and splintered and rimed with saliva. Now, hooking his hands and feet into the rafters, he hangs from all fours, rocking and moaning. He is covered with scabs and sores; his scrotum is bright red and chapped like a cashew fruit.

"He was a quiet little boy," Francisco explains. "Sometimes he would talk to himself—but we all do that, don't we?—and rock quietly."

Sebastão grew up and became a respected citizen, a good provider for his wife and children. One morning fifteen years ago, he entered the forest to go hunting. He was gone for several days, which was not unusual. But when he returned in the night, he skulked into the homestead and, ranting, attempted to chop his family to death with a machete.

"Possession by forest spirits happens all the time here." Senhor Francisco says. "There is nothing you can do. There have always been mapinguarís in these woods, you know. They have only one foot and one eye. They are afraid of water and will never swim across a river. You can tell where they've been because they reek. I spotted one years ago at Seringal União, walking across my estrada, but I saw only its back. Lucky. You have to be careful in these woods. Never look a mapinguarí in the eye, or it will possess you. No one can tell you how it happens, because by then it is too late."

Sebastão has classic symptoms of paranoid schizophrenia. In São Paulo or Rio de Janeiro, he could be treated; a few molecules of medication could transform him. But in the Serra Divisor, possession by a mapinguarí is a terminal affliction.

In a week or two the rains will be continuous, day and night, not like the intermittent storms we've been encountering. These are the last days when it's still possible to burn the slash in the pastures and fields. Tonight our neighbors' fields are in flames and we are enveloped in smoke. The haze above our campfire gathers a cell of light. A few pale birds turn in the white cloud, disoriented, afraid to land, then careen over the dark river. The sky flashes with lightning, brightens uniformly in all directions, as if we were inside a neon tube, as if the land were emitting some sort of static discharge into the air, but there is no accompanying thunder. It is heat lightning, generated somewhere over the horizon, its light muffled and scattered by smoke, the sound of its thunder absorbed by a trillion particles of soot.

All night and into dawn the implacable fires dig into the land and refuse to die. The burning of the Serra Divisor is unlike anything I've ever experienced. It wounds my hope, the hope that has always fed my my science. I've traveled all my life through the tropics and witnessed their slow immolation. I've fled farther and farther into the wilderness to find solace and beauty and to decipher the fabric of creation. Each sunrise, each new morning, was enough to give me hope that there would always be more. But this dawn, entombed in smog, is without redemption.

The smoke enveloping our camp has the stench of burning tires. A tropical forest is, of course, much more than mere wood and leaves. I am smelling boiling saps and latexes, uncountable fricasseed insects, tons of formic acid boiled from the bodies of trillions of ants, simmering lizards and snakes and venom. I am inhaling burnt frog mucus, doomed pollen, roasted nematodes, singed orchids, thrip wings, panicked centipedes, worker termites baked in their earthen domiciles as they gather around their scented queen. I am steeped in denatured flavonoids, terpenes, alkaloids, proteins — all the wonderful stereochemicals crafted for insect antennae and mammal glands, able to trigger ovulation and alter moods—all torn apart, their language made dumb. I am bathed in burnt chlorophyll designed to capture the morning and bring it to earth, to intercept heaven's light and make it into food for the small things.

This is the immolation of Eden. Where do I retreat to now? There is no more wilderness here, no wet garden. Do I take shelter in these puny words, in the bright, toxic screen of my laptop? Do I turn to dogma—santo daíme—take shelter in faith? Invoke the names of saints? I doubt they've noticed. I suspect that the cold universe doesn't care about this little planet or her matricidal children. There must be a billion other living planets with better-behaved offspring.

Wiping the red dust of the holocaust that was once Chiquininim's forest from my glasses, squinting into the troubling light, I see a flush of pale green: a *Trichilia,* a small understory tree in the mahogany family that grows no more than five meters tall. Why this tree survived the fire is a mystery wrought by the chance of ax and flame. It stands alone in a field of toppled giants curled in death. Its trunk is no bigger around than my wrist, an adolescent presenting the season's first new leaves, unfolding to gather photons and rain-borne nutrients, and its first cream-colored flowers, hopeful offerings designed to attract bee senses, but futile now that there are no bees to weave their sinuous paths from flower to flower, to read their messages of scent and color. Now there are no eyes to perceive the gifts of those flowers; there is only the sullen ash.

I walk over to this doomed stripling and in silence behold its bark, not yet pocked by beetles; each of its ten thousand leaves is a cosmos fretted with veins coursing with new sugars. It is sunlight immaderate. Then I take its life—three swift swings of the ax through the soft yellow wood—and fold its pinnate compound leaves and lovely flowers in newspaper, like an undertaker arranging a corpse. I guess I wanted to put it out of its misery, the way one shoots a crippled horse. I don't want to curl up in my hammock tonight and imagine the tree standing alone, bereft of its kind in the abiotic soot. And, too, this is my science; this tree is an important find. The specimens it yields will provide some sort of memorial—like the bust of a minor hero in a forgotten corner of a park. Specialists will take note of its existence, will record that it existed. At least this tree will have a name.

Now the wind has turned northwesterly, away from the Serra. The air is clearing a bit, and the overpowering odor has dissipated. But the air is still unhealthful, stale and painful to inhale. The moon is just shy of full tonight. Earth's captive little planet scurries through yellow layers of mist and cloud, her cold secondhand light making us all restive. Tarzan is tinkering with the fire; I am writing by its light; Arito is sharpening a batch of fishhooks; Geni is plucking a mutum; Pimentel is on the riverbank fishing. Too distracted to take more than a few notes, I lie in my hammock facing east, sleepless. Soon Venus rises, angry and sun-bright, bobbing on the wind-tossed treetops of the dark Serra. I wonder: Did I ask Dona Ausira if there was a Nokini name for "planet"?

Exactly twelve moons ago I made a pilgrimage to western Australia. Earth has traveled completely around the sun since then. The seasons have asymmetrically glazed the antipodes with ice, the Rio Moa has flooded its forest and withdrawn, the fish have grown fat and then skinny. My destination down under was a nondescript place with the aboriginal name of Warrawoona, no more than a heap of coarse-grained quartz marbled by smoky bands of chert, rising from the yellow outback of sun-splintering *Spinifex* grass.

Embedded in the cherts of Warrawoona are layered sheets of marine algae and sediment known as stromatolites. You need a microscope to see the individual plants, so-called dubiofossils, each no more than a simple string of cells. But these algae lived during the Archean epoch, three and a half billion years ago. Contemporaries of the earliest tracings of life on Earth (only a billion years after Earth herself coalesced from the interstellar dust and debris of supernovas), they were present at the first dawn, the first day, the first year of life. I had resolved to see this place where the great journey of my ancestors—and of all the organisms that are my brethren, my food, parasites, and predators—may have begun.

At Warrawoona the horizon seemed to drop off the edge of Earth. The distant hills merged and emerged from a horizon of mirages and wormy heat. A few flattened clouds scattered the eastern horizon, inverting the expanse of land and sky. The only sounds were the insistent wind and the monotonous call of a topknot pigeon, which flew between a gum tree and a low shrub, squeaking with each wing stroke. Silence again. A wedgebill sounded like a distant wind chime.

During the Archean, the hill at Warrawoona was under a shallow sea at the edge of a continent that has since crumpled and moved on tectonic plates all over Earth's face. Conditions on Earth were so different then that the stromatolites may as well have been from another planet. The radiance of the infant sun, heavy in hydrogen but light in helium, was about 30 percent colder than it is today, but Earth's atmosphere stayed within the narrow temperature range of liquid water because it had high levels of carbon dioxide—an early "greenhouse" effect. The days were shorter then—there were about four hundred a year—because the newborn Earth spun on its axis about 20 percent faster than it does today. And the orbital pathway of the moon was ten or twenty thousand kilometers closer to Earth. The shorter days and nearer moon resulted in more frequent and powerful tides than we have now. The layers of stromatolites were formed when sediment stuck to the film of algae with each draw of those tides. Each band represents about a third of a day; every third band represents an Archean dawn.

Why do I think of Warrawoona now, in my hammock under the same moon that drew the short, fast tides over that vagabond continent? Perhaps because I feel that here I am witnessing that journey's end. Today, in this place, even the epiphany of a clear sunrise, the eternal pleasure of awakening, is denied. Will there be many more hematite dawns? A wilderness as vast as the Amazon may never exist again, just as nothing like it ever existed before. Earth will never recover from its loss. Her children will have consumed her very bloom. I can't imagine what the Serra Divisor will be like two hundred years from now, or how these words will sound then. Quaint anecdotes from this brash frontier? A final romantic eulogy to the wilderness?

These are the last days of classic exploration, as numbered as the last remaining patches of wilderness. Science is no less of an adventure now, mind you, but it has become introverted and reductionist. Today a young biologist can spend a lifetime exploring a few nanometers of DNA, and in this microrealm discover pervasive truths that govern blue-green algae, blue whales, and trees sixty meters tall. Of course, DNA is Earth's language. This tropical forest is also wrought of that language, woven into the same fabric of nucleotides, expressed inside uncountable cells.

Above our camp is a spreading angelim tree, sixty meters high, extending its branches just as wide. Chiquininim spared it because it provides shade for his house. The angelim is deciduous and the tree, barren in this dry season, is silhouetted against the clarid new sky. It is not a static wooden scaffold, but an overture to the movements of planet and star. Wrought into its DNA are the instructions for being an Earthling.

The angelim is a gesture to the movements of heaven, a primal gesture, because Earth's star has existed for longer than Earth itself. The tree knows how to push up through the crowded forest, its broad, straight back raising the branches, those boulevards of energy that spread over the canopy and filigree clear space into a million light-absorbing planes. It knows how to project a leaf, as well as the twig and branch that bear it, with just the right economy of material and surface area to capture the light; it knows how to switch on the machinery of photosynthesis in an instant when a mote of light slides past. Its boughs bear the distortions made by vanished neighbors, jogs and ambles that kept them in the light over the years. Some of its branches bear enormous burdens of epiphytes: bromeliads, ferns, orchids, mats of mosses that on a rainy day must weigh tons, and athletic vines that leap to the air to steal the light. They make shadows, pockets of still air, places of small tempest, ideal gaps for silk webs, and kilometers of fissures that are good places for the air-wafted small things that dust the forest to hide. The community of animals that lives on the angelim knows how to mimic a bee's sex pheromone; knows the exact frequency with which a male spider strums the web of his intended so as not to provoke but seduce her. It knows the precise throw weight of a spore in the still air; the route of an entomophagous fungus through a moth's brain; the kingfisher's calculation of the refractive index of water. I could spend a lifetime devoted to understanding this single tree.

Dismembered, the angelim is a few board feet of wood for the sawmill in Cruzeiro do Sul (paneling for the playroom? a child's

coffin?). Its value is a wad of paper *reals* or a trickle of electrons between computerized bank accounts in São Paulo. Burned, its ash
may yield a few kilos of manioc, may get some heroic frontier family through another tough day. Who can fault them? Who would
dare? But these are the reasons why we burn our library on this cold
night. Why we burn our history, our dictionary, our family album.

This is why we immolate the Word.

EPILOGUE

BY THE END of the millennium, the broad beach in front of Cruzeiro do Sul—and all the life and commerce associated with it—had been swept a half kilometer downstream.

By the early 1990s, most of the settlers of Santa Luzia had fled their homesteads and dispersed into the favelas and gold-mining camps of Amazonia, forgotten and dispensable. Dona Cabocla stayed on in the dark forest, cared for by her devoted daughter Gârdie and son Antonio. Theirs were the only viable farms left in the outpost. Although her children had repeatedly invited her to live with them in Cruzeiro do Sul, Dona Cabocla refused to abandon the place she loved, the place of her independence, her República.

On December 7, 1993, Dona Cabocla visited Cruzeiro do Sul to partake in the Catholic festival of Santas Missiones. That evening she suffered a heart attack and died. She was buried in the cemetery on top of the hill.

In 1998 the federal government abandoned Santa Luzia and its utopian dreams of colonizing Amazonia, turning the outpost over to the state government of Acre. The colonists' holdings were consolidated and sold. Two of Dona Cabocla's sons, Antonio and Francisco, bought fazendas there. So did Pimentel. They built a new school, a clinic, a generator, and a water system; the road to Cruzeiro do Sul was paved. By the end of the millennium, Santa Luzia was slowly coming back to life and had once again become a place of hope. But Caboclos—not transplanted eastern folk—live there now. And they have jeito.

The Seringal Aurora is now abandoned, its estradas overgrown. The japiims still nest among the wasps in the silk-cotton tree above the collapsed barracão.

In 2003, on the upper Juruá near the Seringal Tejo, a party of Kampas warriors attacked a neighboring village of Amuacas, their age-old enemies. They killed fifty-two people.

Maciel and his family also abandoned their homestead and the graves of their children on the Rio Moa. Nobody knows where they are.

The Katukina squatters' camp on the Transamazonica has now been formally reserved as tribal land. These days the village has electricity and, at last count, two computers connected by satellite to the Web.

In autumn 1999, a small band of men and women who claimed to be from the Serra Divisor traveled by canoe downstream to Cruzeiro do Sul. When they landed, nobody in the Bairro Flutuante paid much attention to them. The party walked into the FUNAI office and quietly declared that they were Náua. For most of the twentieth century, they said, the Náua had hidden in the Serra Divisor, fleeing disease and exploitation. Now there were about two hundred fifty of them. The Parque Nacional da Serra Divisor, designated to become a wilderness area where people were forbidden to live, would appropriate their lands, they claimed. But their pedigrees proved groundless. The "Náuas," it appeared, were Caboclos from the Seringal Novo Recreio on the Rio Moa, trying to get a parcel of Native American land and stake their own claim on the frontier.

The federal government had been encouraging settlers and speculators to abandon the Serra, promising them cash and new land. The wilderness, for the second time since the conquest, was being depopulated. Chiquininim accepted the government's offer. Fazenda Arizona went feral. Now, more than a decade after leaving his land, he has still not been compensated for it. He makes his living selling snacks from a kiosk on the Praça do Triumfo. "I have joined the ranks of the suffering poor," he told me.

After República was officially declared Indian territory, the Caboclos and whites were required to leave. Because Bolota's sons were

Native American, he was permitted to stay, but it was illegal for him to employ his Nokini neighbors. Without the necessary workers to tend his cattle and crops, Bolota abandoned his fazenda in 1992 and moved to a house in Cruzeiro do Sul. "Why would I want to live like an Indian?" he complained. Over the course of the next year, the Nokini ate all of Bolota's sugar cane and cassava, leaving none left to sell. When the sugar mill broke, it was not repaired.

By the time she died in 2003, Dona María had delivered 2,850 babies. A team from Rede Globo, the national television channel, traveled to República to film a special about her legendary midwifery. The child she treated for snakebite with Dona Ausira's recipe of cumarú cheiroso survived, but her arm has atrophied.

The lone *Trichilia* I collected at Fazenda Arizona proved to be a widespread but locally rare species, *T. solitudinus*. It has never again been recorded in the Serra Divisor.

Dona Ausira died in 1995, and at that instant the Nokini language also perished.

𝕊 *Acknowledgments*

I wish to express my profound gratitude to Ghillean T. Prance for inviting me on my first botanical expedition to the Brazilian Amazon in 1974 and for encouraging and supporting my research there for over a decade, and to William Balée for opening my eyes to subtle yet abiding issues of Amazonian environmental history. From 1984 to 1993, I made five expeditions to the upper Rio Juruá, Rio Moa, and Rio Azul in far western Acre under the auspices of the New York Botanical Garden, the Instituto Nacional de Pesquisas da Amazônia, the Museu Paraense Emilio Goeldi, and Grinnell College. This book tells the story of one of those expeditions, but it derives from all of them. The names of only a few of the people who were my companions, teachers, and guides appear in this account. Instead, I have tried to tell the story of local people: Arito Rosas, Jr., Dionesio Coelho, Francisco Barbosa Derzi, Geni Japiim Derzi, Francisco Pimentel da Costa, and Zecau Breju. My conversations with these people are transcribed from memory, and are not verbatim. Of course, I am inexpressibly grateful to the others on those expeditions: William Balée, Edimar Ferreira Batista, Douglas Daly, Carlos Alberto Cid Ferreira, Dawn Frame, Nilton Tadeu Garcia,

Jacques Jangoux, Ellen McCallie, Scott Mori, Jason Ney, Francis Nishida, John Pruski, Mick Richardson, Edileuza Sette Silva, José Lima dos Santos, Judy Stone, Jay Walker, and David Williams. I can never repay the gifts of friendship and discovery that they have brought to my life.

The research described in this book was supported by grants from the Henry Luce Foundation, the National Science Foundation, Brazil's Conselho Nacional de Pesquisas, and the Jesse Smith Noyes Foundation. I am honored to acknowledge a fellowship from the John Simon Guggenheim Foundation that sustained both my field work and the writing phase of this project. I thank William Balée for reviewing the manuscript and providing many useful comments; Richard Chandless for granting permission to quote passages from the unpublished diary of his great-uncle, William Chandless; and Leidimar Rodrigues de Andrade for verifying the spelling of Portuguese words. As always, Harry Foster and Peg Anderson, my patient editors at Houghton Mifflin, made innumerable valuable improvements.

Finally, I thank my wife, Karen Lowell, and daughter, Tatiana Lowell-Campbell, who shared many of the experiences described in this book, as well as the adventure of writing it. Our journeys together have been the greatest beauty of my life.

◈ Notes

Prologue

page 10 *There is a wooded:* Seymour, p. 367. The goats Homer refers to are mountain goats, not the ancestors of the domesticated goat.

11 *Mythical* mapinguarí: In a pioneering effort of xenobiology, Dr. David Oren, a zoologist at Belém's Museu Paraense Emilio Goeldi, led two expeditions to forests in Acre to collect proof that the *mapinguarí* is a ground sloth. As of this writing the results have been inconclusive (personal communication).

2. River of Light

38 *It is a male:* In jacana society, the father builds the nest and rears the chicks; the promiscuous female lays her eggs in her mate's nest and departs. The nests are insubstantial, made of a few slipshod floating reeds and lily pads. When the chicks grow too large to be inconspicuous, the male carries them tucked under his wings; he becomes, in effect, a mobile nest.

40 *Even though we are only:* How do I know I am breathing Atlantic vapor? After all, the raindrops here look the same as any others. The raindrops carry the secret information of their source, and the discovery of this fact, by Eneas Salati and his colleagues at the National Amazon Research Institute, was one of the greatest but least celebrated insights of twentieth-century science. The researchers derived this information in two ingenious ways, with experiments that used continents and rivers as their benchtops. First, hydrologists measured the amount of rainfall in a representative area of Amazonia and subtracted from that value the amount of river runoff. Assuming that the forest is in hydrological equilibrium (that is, the amount of water entering the system equals the amount leaving it), then the total rainfall minus the

runoff must equal the amount returned to the atmosphere through evaporation and transpiration.

Second, they measured the ratios of two common isotopes of oxygen, O^{16} and the much rarer O^{18}, in the atmospheric water vapor at eight points along a 3,000-kilometer-long transect from the Atlantic Ocean to Benjamin Constant, near the Peruvian border. With each successive assimilation by the forest, the ratio of O^{16} to O^{18} in the rainwater molecules increased. In other words, at each step along the way the rainwater became isotopically lighter. This is because water molecules containing O^{18} are 11 percent heavier than those containing O^{16} and therefore are less likely to vaporize, whether from a leaf's surface or through its stomata. Given that almost all atmospheric water over the Amazon comes from the Atlantic Ocean, the rate of attenuation of O^{18} in water vapor can be calibrated to reveal how many times it has fallen as rain and has been recycled back into the atmosphere by the forest.

The startling result was that by the time the atmospheric water vapor reached Manaus, 1,500 kilometers from the Atlantic, it had been recycled as many as fifteen times; by the time it reached Benjamin Constant, about thirty times. Each passage takes about five and a half days; therefore, the water vapor I am breathing today on the Rio Moa blew in from the Atlantic five or six months ago and has been incorporated into the very substance of this vast forest.

In this manner, the westward passage of the atmospheric water vapor is slowed down and the capacity of the land and its mantle of vegetation to hold water is greatly enhanced. If the forest is cut down, rainfall will decrease substantially, not only over the Amazon itself but also in the areas to the south, including Brazil's grain-belt states of São Paulo, Paraná, Santa Catarina, and Rio Grande do Sul, where seasonal winds shunt Amazonian atmospheric water vapor. When first published in the mid-1980s, this information created a stir in the agricultural and economic sectors of Brazil, because it showed for the first time that Amazonian deforestation could decrease agricultural production hundreds of kilometers to the south. For the first time, the consequences of Amazonian deforestation on the pocketbook of the nation were clear. Salati et al. 1983; Salati and Vose, 1984.

42 *all the same cohort:* A single, synchronous flowering and seed-bearing (known as monocarpy), followed by death, is common in bamboos in both the Old World and the New. The periodicity varies on the scale of decades, depending on the species. The simultaneous die-offs in the western Amazon are so extensive that they have been detected by satellites. This phenology is believed to be an adaptation to predation, a mechanism to catch the bamboo-eating creatures off guard and cause their populations to crash. For example, the mass flowering and die-off of bamboos in Sichuan, China, is one of the principal threats to the long-term survival of the giant panda, which feeds exclusively on bamboos, in the wild.

45 *the nutritional virtues:* A fruit, whether dispersed by bird, mammal, or fish, must never fully satisfy its vector. If it did, then there would be no motivation for the vector to move on and distribute the seeds.

They grow fat: Seventy percent of the food produced in Amazonia comes from inundated areas. The protein potentially harvestable from fish in the várzea forests is ten to fifteen times greater than what could be produced by beef. If this forest were cut down to make cattle ranches, it would never yield the amount of protein that the intact forest does. For one thing, cattle, being warm-blooded, must expend calories to maintain their body temperature, energy that otherwise would go into growth. Fish, being cold-blooded, are not as energetically wasteful. Smith 1979.

3. An Expedition of Poets

49 *Toward what inevitable:* Translation by the author of Camões, "O Desencanto do Eldorado," *Os Lusíades,* Canto IV

A que novos desastres determinadas
De levar êstes reinos e esta gente?
Que perigos, que mortes lhe destinas
Debaixo de algum nome preeminente?
Que promessas de reinos e de minas
D'ouro que lhe farás tão fàcilmente?
Que famas lhe prometerás? Que histórias?
Que triunfos? Que palmas? Que vitórias?

53 *The story of rubber:* Amazon rubber should not be confused with the product used by the Maya and Aztecs in Mesoamerica. Every Classic Maya city had a ball court where teams of men (sometimes, it is said, prisoners of war) ceremonially battled each other in a game that was a precursor of soccer. Just as in the modern game, players were allowed to move the ball only with their feet, legs, torsos, and heads. By the time the Spanish arrived in Central America, most of the Maya cities were abandoned and their ball courts overgrown with forest, but the game had been adopted by the Aztecs.

Some Aztec balls, which weighed up to twenty kilos, were presented to the court of King Carlos V in 1524, only three years after the Spanish conquest of Montezuma. That year Andrea Navagiero, the Italian ambassador to Spain, was invited to watch a game of "tennis" played by a team of Aztec prisoners. It may have been Europe's first soccer game. Navagiero described the ball as being made "of some kind of very light wood . . . which bounced with extreme ease."

The rubber described by Navagiero was not the product that became an article of international commerce over the next five hundred years, made from the latex of South American trees in the genus *Hevea* (most commonly the Brazilian rubber tree, *H. braziliensis*). The latex used by the Maya and Aztecs was probably from the caucho tree (*Castilloa* sp.), in the fig family. Although *Castilloa* is found in Amazonia and is sometimes substituted for *Hevea,* it is not the latex of choice for industrial rubber. Moreover, unlike *Hevea,* the *Castilloa* tree must be killed for its latex to be harvested — hardly conducive to a long-term enterprise.

The Aztecs' elastic material didn't catch on in Spain. At most, it became an expensive toy for the Iberian aristocracy and their privileged children. Ironically, one reason for the lack of interest may have been that it was so coveted. The Spanish censors, who controlled all published information on all products of potential commercial value from the New World, stifled the news of its discovery, and except for Navagiero's descriptions, no information on the wonderful new product went beyond Iberia for another two hundred years.

54 *believed to be banana leaves:* Bananas are Old World plants that were just being introduced to the Americas in 1736. The leaves may have been those of a New World relative of the banana in the genus *Heliconia.*

"In the forests": Coates 1987, pp. 6–7. Coates's translation has been slightly modified by the author.

"Indians call it": The Spanish word for rubber, *caucho,* is derived from the Quechua *caoutchouc,* which means "weeping wood."

55 *"The Omáguas make":* a Condamine, 1747, quoted in von Hagen 1948, p. 127.

57 *"The labor of extracting":* Hay 1870, p. 294.

The extraordinary price": Spruce 1908, p. 233.

58 *"They overcrowded":* Cunha 1986, p. 55.

62 *named tributaries:* They are the Rios Ipixuna, Moa, Arrependido, Liberdade, Tauaré, Gregório, Acurauá, Tarauacá, Muru, Envira, Igapó Araça, Jurupari, Pauini, Moaco, Macauã, Antimari, Rozinho do Amdirá, Rola, Acre, Edinari, Iquiri, Ituxi, and Abuná. The rivers are only a few kilometers apart. To make the 510-kilometer journey from Cruzeiro do Sul to Rio Branco, the territorial capital, would require exhausting portages across repeating bands of terra firme, interspersed by river and várzea—a journey that no rational person would attempt until 1970, when the Transamazonica was bulldozed across the same route.

64 *"It is just":* Tocantins 1979, vol 1, p. 275. Translation is by the author.

65 *we still have the receipt:* The receipt, which is archived in the Archaeological Institute of Pernambuco, is extracted from Tocantins (1961) vol. 1, p. 280. The monetary units are reis. Translation is by the author.

68 *Estiraõ dos Náuas:* Meander of the Náuas.

69 *"the true story":* Hepper 1982, p. 131.

"cajoling a friend": Wickham 1872.

Britain had broken: The acquisition of rubber came at the end of this marvelous exchange, which had begun more than a century earlier with the founding of Kew in 1773. Kew Gardens soon became the pioneer of Western botanical exploration and experimental horticulture. Its first director, Joseph Banks, cut his teeth as a botanist on Captain James Cook's circumnavigation of the Southern Hemisphere. Banks inaugurated the tradition of dispatching collectors from Kew all over the world, beginning in 1787 when he sent horticulturist David Nelson on the HMS *Bounty* to take breadfruit plants from Polynesia to the West Indies. Nelson failed, perishing eleven weeks after having been set adrift by mutineers in an open lifeboat with Captain William Bligh, but two years later his successors, Christopher Smith and James Wiles (also gardeners at Kew), succeeded in introducing the species to St. Vincent. Banks was followed as director by the father-and-son dynasty of William and Joseph Hooker, who for most of the 1800s oversaw the importation and transplantation of thousands of crop varieties, from giant tropical trees to fungi, and in 1848 founded the world's first Museum of Economic Botany. The Hookers, unnoticed in history books, transformed this planet—its mantle of vegetation, its land use, and economy—more than any Napoleon ever could. If one takes an airplane anywhere in the tropics and looks out the window, one sees the results of the vision and labor of these two men.

During the reign of the Hookers, London became the world center of biological imperialism, this transplantation and experimentation with the marvelous living varieties discovered in new lands. It was a heady time for plant explorers. The newly found biological diversity captivated Victorian society; the queen herself ordered that the fastest clipper ships and carriages rush fresh tropical fruits from India to her table. Naturalist-collectors such as Henry Walter Bates, Alfred Russel Wallace, and Charles Darwin returned from collecting forays in the tropics to restructure science and philosophy—in short, to realign our species' comprehension of its place in the universe.

4. River of Hunger

80 *Silvery retinas:* The nocturnal animals without a tapetum are as interesting as the norm. Sloths' eyes don't reflect at all. Nor do bats' eyes; their sonar precludes any need for night vision.

84 *By sliding:* By controlling the nest temperature, the mother jacaré may also be choosing the sex of her offspring. Gender in crocodilians is not genetically predetermined, as it is in mammals, but occurs after the egg is fertilized. It is a function of the temperature of incubation: if it is higher than about 40°C, the babies will be mostly males; if it is cooler, they will be females. Whether the mother responds to something in the environment and adjusts the sex ratio of her babies accordingly is unknown. In theory, at least, maternal choice could be adaptive. For example, if the caimans' population density is low and there is a surplus of suitable nesting places, it might be to her advantage for most of her babies to be female; if the river is already heavily populated with females, it might be better to make male babies.

85 *The value of a single:* Only a wet hide, salted and kept moist, could command such a high price. Dry hides, lightweight and easy to transport, were less pliable and resisted being worked into handbags and shoes. Salted hides, of course, were very heavy, and Arito and his team could not carry very many at a time to market in their small boats.

5. What the Bees Said

90 *"Bem ti ve":* "Good to see you; good to see you!"
"Todos velhos": "All old men are ugly, all old men are ugly, except for my grandfather, except for my grandfather."

94 *beans, papaya, and bananas:* Although the banana is a native of the Old World, it has been adopted by New World tropical agricultural economies, cultivated even by the most isolated and remote Amazonian tribes.

99 *terras roxas are good:* Moreover, the terra roxa has a distinct and characteristic vegetation. It is viny and populated with trees such as the babassu palm, which survives fire, and *Alexa imperatricus,* an indictor of a very old disturbance, the last stage of the forest healing itself. Viny forest comprises about 11 percent of the terra firme in eastern Amazonia, and until recently it was thought to be a natural phenomenon. Only in the last decades of the twentieth century did ecologists perceive the obvious: that it is a signature of human cultivation on a massive scale (Smith 1980; Balée and Campbell 1989).

6. On the Line

106 *In fact, the lowland:* For a comparison of soil nutrient levels in adjacent stands of várzea vs. terra firme forest in the margin of the Rio Juruá in Acre, see Campbell, Stone, and Rosas 1992.

108 *Most of the trees enlist:* Mycorrhizae are almost ubiquitous root commensals, found not only in tropical forests but in temperate forest trees, epiphytes, and most domesticated crops. Interestingly, sea grasses do not have mycorrhizae.

114 *The cecropias:* Cecropias are lazy. Instead of manufacturing complex, expensive poisons to repel or kill browsing insects, they use the services of stinging ants to do the job. In return they provide the ants with shelter in their hollow stems and cheap but satisfying sugars and oils.

Recent studies: Young and Perocha 1994.

116 *One could argue:* Some anthropologists argue that abandoned swiddens actually increase the carrying capacity of the rainforest for humans, because the secondary forest supports greater densities of game animals—tapirs, peccaries, brocket deer—than the primary forest. Furthermore, the abandoned swiddens often have higher densities of fruit trees and other useful plants encouraged by humans during their brief period of occupation. For a provocative discussion of this interesting idea in the context of the Ka'apor tribe of eastern Amazonia, see Balée 1994.

Pre-Columbian swiddens: See Denevan 1992, 1996.

117 *"In the equable":* Wallace 1891, p. 268.

120 {*Is the frequency:* This concept, known as the "intermediate disturbance hypothesis," was independently derived by Joseph Connell (1978) and Michael Huston; for a thorough review see Huston 1994. One of the most provocative theories in modern ecological theory, the intermediate disturbance hypothesis states that competitive exclusion takes longer in environments that are subjected to moderate levels of disruption. The middle level of disturbance is the key: too much tends to wipe out species indiscriminately; too little permits the most competitively proficient species to overwhelm and push aside the others.

121 *"One day a smart":* Bates 1854, p. 272.

7. The Naming of Parts

132 *Tarzan rolls:* Envíra, in the custard apple family (Annonaceae), can be used as a natural compass. In western Acre (but not necessarily in other parts of Amazonia), bark growing on the sunny eastern side of an envíra trunk peels in thinner strips than bark from the western side (which is toughened by exposure to storms that descend down the Andes).

133 *She plops to the ground:* Later she will take the paralyzed prey to her burrow, dug under a fallen leaf somewhere in the terra firme, and deposit a single egg on it. After the egg hatches, the wasp maggot will slowly devour the still-living spider. It will eat the expendable organs, like legs and antennae first, then, just before emerging, the vital ones like the heart, never quite killing its host until the last minute so that it continues to have a supply of fresh meat.

A bico de argulha: The bluish-fronted jacamar (*Galbula cyanescens*) replaces the rufous-tailed jacamar (*G. ruficauda*) in the western Amazon. Rivers pose a barrier to the dispersal of these birds, and the ranges of *Galbula* species appear to be delineated by their courses.

134 *Studies of decomposing fruits:* Researchers studying the amapá tree (*Parahancornia amapa*), of eastern Amazonia (see Morais et al. 1995), have identified thirty-four species of yeasts growing on the fruits; as is inevitable in Amazonia, one species—apparently a specialist on amapá—was new to science. Two of the yeast species practiced allelopathy—that is, chemical warfare, producing toxic compounds targeted to kill competing species.

The fruit-fly maggots: In the eastern Amazon amapá fruits are colonized by flies in the *Drosophila willistoni* complex, a suite of three closely related subspecies that (according to the fruit fly taxonomists) have not yet differentiated into discrete species. Each subspecies of *D. willistoni* lays its eggs during a particular stage of decomposition, thereby minimizing competition with the others.

137 *Fast-growing species:* Since the 1990s a series of papers by Phillips and colleagues (Malhi and Phillips 2004) have claimed that the turnover rate of tropical forests (that is, the death of existing trees and the recruitment of new ones) had been accelerated by anthropogenic enrichment of the atmosphere by carbon dioxide (the so-called greenhouse effect). Although that may be the case, the studies were based on quantitative inventories of tropical forests such as ours on the Rio Azul. The problem with these studies is that the accelerated turnover rates may have been the result of repeated visits by teams of researchers. The very act of conducting the experiments may have invalidated the results.

139 *About a third have appeared:* At least fifteen of the *Inga* species, and very likely the majority of them, came into existence through polyploidy, meaning that meiotic anomalies caused a sudden multiplication of chromosome numbers. The base chromosome number for *Inga* is thirteen. Of the seventeen species of *Inga* for which there are chromosome counts, thirteen are diploids (having twenty-six chromosomes, twice the base number) and two are tetraploids (having fifty-two chromosomes, four times the base number). Polyploidy is a speciation

process that can take place in a single generation; for all practical purposes it is instantaneous.

139 *This time-place is known:* The eminent evolutionary biologist G. Evelyn Hutchinson, frustrated by his attempts to characterize "niche," eventually conceded that it was essentially impossible to define, calling it a "polydimensional hypervolume."

140 *"Young shoots":* Pennington 1997, p. 771.

142 *Pennington's monograph:* His *Inga* study was funded by the Overseas Development Administration of the United Kingdom, not because of the evolutionary significance of the genus, but because, unique among Neotropical legumes, *Inga* has the ability to colonize the senile, acidic soils that result from large-scale deforestation. Better yet, because it is a legume, *Inga* can restore nitrogen to those soils.

144 *countless names for God:* Arthur C. Clarke's wonderful short story "The Nine Billion Names of God" (1953) was based on this theme.

8. The Gift

154 *Pit-viper venom:* The makeup of the venom is determined by that species' hunting strategy. Ground-dwelling fer-de-lances produce a venom rich in hemolytic proteins. They strike and withdraw, confident that they can easily locate the corpse by following its heat trail. But tree-dwelling pit vipers that feed on birds can't allow their prey to fly away and die, so their venom is rich in paralyzing neurotoxins. They strike and hold on.

9. River of Terns

159 *"[These] provinces":* Leandro de la Cruz is quoted in Porro (1994), p. 79. The translation is Porro's.

Rios Madeira, Purus: These are the modern spellings of the river names. On the map they are Purus, Coary, Jepé, Jurua, and Yantay.

160 *a chronology of the maps:* 1821, *Nova Carta do Brasil e da America Portuguesa.* 1873, *Carta Imperio do Brasil Reduzida no Archivo Militar em conformidade da Publicada pelo Coronel Conrado Jacob de Niemeyer em 1846 e das especiaes das fronteiras com os Estados Limitrophes organsadas ultimamente pelo conselheiro Duarte da Fonte Ribeiro, Rio de Janeiro.* 1892, *Carta da Republica das Estados Unidos do Brasil com a designação das ferrovias, Rios Navegaueis, colonias, engenhos centrales, Linhas Telegraphicas, e de nevegação à vapor. Organisado em 1833 e rectificada por ordem do cidadão Ministro da Industria, Viação e Obras Publicas. Dr. Innocencio Serzedello Corrêa.* 1903, *Mappa Mostrando a Nova Fronteira entre o Brasil e a Bolivia na Região Amazonica.* 1904, *The Acre Territory and the Caoutchouc Region of South Western Amazonia, showing the frontiers according to the treaty of 1903 between Brazil and Bolivia, Geographical Journal.* 1906, *Mapa del Alto Yurua y Alto Purus que comprende las ultimas exploraciónes y estudios verificados desde 1900 hasta 1906.*

161 *among the last places:* In fact, it was not until the 1970s, with the advent of satellite imagery, that some Amazonian tributaries were discovered.

162 *The epidemic of 1660:* Why were the Native Americans so susceptible to European diseases? And why didn't they transmit their own diseases to the invading conquistadors? The answer to the first question may be the Europeans' septic proximity to their domesticated animals, a relationship that had begun thousands of years before the conquest of the Americas. The domestication of animals presented opportunities for zoonoses to leap to humans: smallpox and measles are derived from diseases of cattle, influenza from fowl and swine, plague from rats. As to the second question, Native Americans, from Alaska to Patagonia, had only two domesticated mammals, the dog and the guinea pig; a semidomesticated mammal, the llama; and one fowl, the turkey. Epidemiological innocents, they shared few contagions of animal origin and had few to inflict on the Europeans. See Diamond 1986, pp. 195–214.

163 *"In former days":* Fritz 1922, p. 64.

"Here are collected": La Condamine quoted in Ross 1978, who in turn cites Earl Hanson (1967), *South From the Spanish Main,* New York: Delacorte.

164 *In 1639 Cristobal de Acuña:* Noting the northerly flow of the Juruá, Acuña named it Rio Cusco, assuming (incorrectly) that it joined the Río Urubamba, the steep-walled Andean torrent on which the Inca aerie of Machu Pichu is situated. Acuña, "A New Discovery," cited in Ross 1978.

165 *"the best sarsaparilla":* Ibid., pp. 75, 77.

167 *"All the christianized":* Herndon 1854, pp. 256–58.

"In the vicinity": Edwards 1847.

168 *"I saw here":* Bates 1876, p. 257.

"Great mortality": Ibid., p. 259.

"We took": Ibid., p. 2.

10. A Land of Ghosts

173 *we must use ceramics:* The transition from the stone-working of the Andean tribes to the making of clay objects in the pebble-free Amazonian lowlands has led to the belief that the traditions of the advanced civilizations of the Andes (and the tropical Pacific coast of South America) permeated into the lowland Amazon, where craftsmanship and sophistication were slowly lost. But archaeological evidence of widespread lowland ceramic traditions, from the Ucayalí to the Caribbean, imply that the trend may have been in reverse, that the Andean traditions were secondarily derived from lowland ones. See Lathrup 1970; Roosevelt 1994.

174 *"an empty social landscape":* Nugent 1994.

"I certify": Monteiro de Noronha, quoted in Reis 1931, p. 160. (Translation by the author.)

175 *"the Juruá has been":* Monteiro de Noronha.

"bastard brother": João Augusto was a colonizer of the central square in Pará (now Belém).

"My father": Quoted in Monteiro de Noronha p. 167.

176 *"The youngest daughter":* Ibid., p. 165.

177 *"Black and white waters":* All quotes are from Chandless 1867 or Chandless 1869. His unpublished journal (1867) is used by permission of Richard Chandless.

181 *"the vast, immaculate beaches":* All quotes from Tastevin are from his numerous papers (1920–1928) or those by Rivet and Tastevin (1921–1938), which were translated from French by the author.

"The black waters": He continued: "Is this prolonged maceration of vegetable detritus that which poisons the waters and gives them their amber color? Could be!" He was right, of course. The black color comes from tannic and humic acids released by decaying leaves.

182 *"voluntary sterility":* The mechanism for their sterility is unknown. Tastevin noted that they used a plant named *imi-rau* ("blood medicine"). "A woman doesn't want more than one or two children, because she is very depressed and her husband is desperate."

11. Angelim

188 *"Suddenly the sun rose":* Translation by Kullberg 1993.

191 *The flock first settled:* Today the trail from Cachoeira Formosa is one of the principal cocaine smugglers' routes between Peru and Brazil. It is known as the Caminho do Diablo—the Devil's Walkway.

196 *ebullient blond bitch:* Xuxa is named for the blond, coquettish, buxom hostess of a popular Brazilian children's television program.

207 *so-called dubiofossils:* Since their discovery and first description by William Schopf, there has been an enduring controversy as to whether the "dubiofossils" of Warrawoona are the signature of Earth's earliest life forms or the result of abiotic processes. I accept Schopf's opinion that they are fossilized algae. In any event, it is widely accepted that life on Earth was abundant during the Archean epoch.

𝔖 Species Descriptions and Glossary of Portuguese Words

Abiu Fruit-bearing trees in the genus *Chrysophyllum* (Sapotaceae).

Acaba da noite ("wasp of the night"). Parasol wasp, *Apoica pollens* (Vespidae), the principal nocturnal wasp in Amazonia.

Acaba da igreja ("wasp of the church"). Mud-dauber wasp, a hunter of caterpillars, in the genus *Polistes* (Vespidae).

Acaba tatu ("armadillo wasp"). Drumming wasp, *Synoeca* species (Vespidae).

Açaí Palm, *Euterpe oleracea* (Arecaceae), with purple fruits that taste like a mix of pears and Kaopectate.

Acari Any of several species of armored catfish in the genera *Hemiancistrus, Pteroplichthys,* and *Plecostomus* (Loricariidae) that hide in the boles of *acariquara* when the water level is high.

Acariquara ("fish place"). Tree, *Minquartia guianensis* (Olacaceae).

Achilles morpho Butterfly, *Morpho achilleana* (Nymphalidae).

Agouti *Dasyprocta fuliginosa* (Dasyproctidae), lowland relative of the Andean Guinea pig.

Agrovila Planned settlement on the Transamazonica.

Água estapada Anoxic, still water, usually in an isolated oxbow lake.

Águia-pescadora Osprey, *Pandion haliaetus* (Pandionidae).

Aldeia Missionary village, usually Jesuit in origin.

Amarelinho Tree, genus unknown, in the family Apocynaceae.

Amazon kingfisher *Chloroceryle amazona* (Alcedinidae).

Ameiva Lizard in the genus *Ameiva* (Teiidae).

Andorinha-do-mar-preta Brown noddy tern, *Anous stolidus* (Laridae).

Angelim Emergent tree in the genus *Parkia* (Fabaceae), that grows to sixty meters.

Anil Indigo.

Aninga do igapó Arum, *Montrichardia* (Araceae).

Anta Brazilian tapir, *Tapirus terrestris* (Tapiridae).

Aracuã Variable (speckled) chachalaca, *Ortalis guttata* (Cracidae).

Aracu-pinima Fish with black and golden vertical stripes, *Leporinus affinis* (Characineae).

Arapuá Stingless bee in the genus *Trigona* (Meliponinae), about the size of a housefly, that ranges throughout the New World tropics. These bees build subterranean nests, which they enter through a wattled mud siphon. Before the introduction of high-yielding African honeybees to the Americas, native Central and South Americans domesticated meliponids for their honey; the Maya made a mead from it.

Arara-canga Scarlet macaw, *Ara macao* (Psittacidae).

Arpão Two-barbed harpoon used to hunt caimans.

Arpoador Harpoonist.

Aruanã *Osteoglossum bichirrhosum,* one of two Amazonian species in the family of bony-tongued fishes (Osteoglossidae). The other is the *pirarucú*.

Assacú ("burning ass"). A tree, *Hura crepitans* (Euphorbiaceae), with caustic sap. The name refers to the dermatitis that results from casually sitting on the tree.

Aviação Grubstake provided to a rubber-tapper by his *patrão*.

Bacurão Nightjar, goatsucker. Nocturnal birds of several genera (Caprimulgidae).

Bacuri Tree in the genus *Rheedia* (Clusiaceae) that has an edible fruit.

Bairro Neighborhood; slum.

Bairro Flutuante ("floating neighborhood"). Cruzeiro do Sul's waterfront neighborhood built on pontoons and stilts above the mud flats.

Bananeira Showy herb, *Heliconia* species (Heliconiaceae), with red or orange inflorescences.

Barracão ("barracks"). Rubber tappers' dormitory.

Batedor Professional Indian-hunter.

Beija-flor-de-banda-branca ("flower-kisser-of-the-white-flag"). Versicolored emerald hummingbird, *Amazilia versicolor* (Trochilidae).

Bem-te-vi Tyrant kingbird, great kiskadee, *Pitangus sulphuratus* (Tyrannidae).

Bico de argulha ("needle-nose"). Bluish-fronted jacamar, *Galbula cyanescens*.

Black skimmer *Rynchops niger* (Rhynchopidae).

Black vulture *Coragyps atratus* (Cathartidae).

Black-tailed trogon *Trogon melanurus* (Trogonidae).

Black-winged stilt *Himantopus himantopus* (Recurvirostridae).

Bladderwort *Utricularia longifolia* (Lentibulariaceae).

Bola Ball of smoked, coagulated rubber latex.

Botfly Fly with parasitic maggots in the genus *Philornis* (muscidae).

Botinho "Talking catfish," *Hassar wilderi* (Pimelodidae), which makes a squeaking sound by stridulating its swim bladder with a bone.

Bôto vermelho ("red dolphin"), *Inia goeffriensis* (Platanistidae), one of two species of small whales found in Amazonia. Today this relictual family is relegated to the Yangtse, Indus, Ganges, and Amazon/Orinoco rivers. *Bôtos,* which grow to two meters, show their bulbous heads and blunt dorsal fin when they breach, but no flukes. There have been numerous trustworthy reports of pods of *bôtos* and *tucuxís* (dolphins) fishing cooperatively.

Brown noddy tern *Anous stolidus* (Laridae).

Buriti Common fan-leaved palm, *Mauritia flexuosa* (Arecaceae), an indicator of the floodplain.

Caapi Vine, *Banisteriopsis caapi* (Malphigiaceae), whose bark is used to make *santo daimé. Caapi* bark is boiled with the leaves of *café brava* (literally "wild coffee"), any of a half dozen species of *Psychotria* (Rubiaceae).

Caboclo Amazonian of mixed descent but culturally Native American.

Cachaça Cheap white rum, *águardiente.*

Cachoeira Waterfall; rapids.

Caiçuma Native American ceremony in which *chicha* is drunk.

Calliandra Genus of shrubs and trees in the family Caesalpiniaceae.

Campo Open field.

Canelão de velho ("shinbone of an old man"). A tree, *Aniba canelilla* (Lauraceae).

Canoeiro Tree frog, *Hyla* species (Hylidae).

Capanga Bodyguard.

Capim Grass, *Paspalum conjugatum* (Poaceae).

Caracará Crested caracara, *Polyborus plancus* (Falconidae).

Caranguejeira Tarantula, probably *Theraphosa* species (Theraphosidae).

Carapanaúba Tall tree in the genus *Aspidosperma* (Apocynaceae) with a sulcated trunk.

Caucho Coagulated latex of a fig tree, *Castilla ulei* (Moraceae); it is a second-rate substitute for true Brazilian rubber.

Cearense Person from Ceará, a state in northeastern Brazil.

Cedro Tree, *Cedrela odorata* (Meliaceae), related to the mahoganies, characterized by deep sulcations in its trunk.

Channel-billed toucan *Ramphastos vitellinus* v. *culminatus* (Ramphastidae).

Chicha Alcoholic drink made of fermented manioc.

Chiming wedgebill Bird, *Psophodes cristatus* (Cinclosomatidae).

Churrasco Barbecue of mixed meat kebabs roasted on a sword or spit, borrowed from the gauchos of southern Brazil, Uruguay, and Argentina.

Cigana ("gypsy"). Hoatzin, *Opisthocomus hoazin* (Opisthocomidae), a leaf-eating bird with clawed wings.

Cinzento Yellow-crowned tyrannulet, *Tyrannulus elatus* (Tyrannidae), a flycatcher.

Coatimundi *Nasua nasua* (Procyonidae), a relative of the North and Central American raccoons. Coatis, highly social animals, forage in packs of as many as twenty, led by an experienced matriarch. Highly vocal, they snuffle, squeak, and whistle.

Coca Shrub in the genus *Erythroxylum* (Erythroxylaceae); the leaves are the source of coca paste, which is refined into cocaine.

Common potoo Large nocturnal bird, *Nyctibius griseus* (Nyctibiidae).

Copaíba Aromatic oil, used as a topical medicine throughout Amazonia, from the tree *Copaifera multijuga* (Caesalpinioideae).

Coreira Hunting party intended to kill Native Americans.

Coringa Crude tobacco that is coiled into ropes and dried. A poultice of it is commonly used to kill botfly larvae.

Coró Spiny tree rat, *Mesomys hispidus* (Echinmyidae).

Corta-água Black skimmer, *Rynchops niger* (Rhynchopidae), a black and white bird with a red bill.

Cortiça Balsa wood float used in hunting caiman.

Crente ("believer"). A member of any of a variety of fundamentalist Protestant sects in Amazonia.

Cuidado Be careful.

Cumarú cheiroso Tree in the genus *Dipteryx* (Caesalpiniaceae).

Curassow Large bird, *Crax* species (Cracidae), esteemed as food.

Curimatã Fish in the genus *Prochilodus* (Curimatidae).

Cutia Agouti, *Dasyprocta fuliginosa* (Dasyproctidae).

Diária Daily wage.

Doenca infantil Childhood sickness.

Dog-faced bat *Molossops* species (Molossidae).

Doido Idiotic.

Dragon tree *Dracaena draco* (Dracaenaceae), native to the Canary Islands.

Drogas do sertão ("drugs of the outback"). Resins, latexes, barks, balms, honey, and other products extracted from the forest.

Egg-eating snake *Drymarchon* species (Colubridae).

Ejercito da Boracha Army of Rubber.

Electric eel *Electrophorus electricus* (Electrophoridae).

Entrada ("entry"). Commercial expedition to kidnap slaves.

Envíra Tree in the custard apple family (Annonaceae).

Estrada ("street"). Rubber tapper's trail.

Estupidamente gelada ("stupidly frozen"). A recent idiom, always used to describe beer.

Farinha Flour made from the rhizome of bitter manioc (*Manihot esculenta* (Euphorbiaceae).

Fava bulacha ("biscuit bean"), *Vatairea guianensis* (Papilionoideae).

Fava de tambaqui *Macrolobium acaciifolium* (Caesalpiniaceae).

Favela Urban shantytown, slum.

Fer-de-lance *Bothrops atrox* (Viperidae).

Filho Son.

Fiscalização ("fiscalization"). Accounting.

Fish-eating bat *Noctilio* species (Noctilionidae).

Flor de Junho ("flower of June"). A seasonal metaphor from Portugal that also works in western Amazonia, since June is the beginning of the dry season, when most plants flower.

Flor de mulher ("flower of woman"), *Clitoria* species (Fabaceae). The name refers to the striking resemblance of the curling petals to the folds and curves of a woman's sex. Carolus Linnaeus, the Swedish founder of modern taxonomy, noted this resemblance and assigned the genus name.

Friagem Vernal storm generated in Antarctica, bringing cold wind and rain.

Futebol Soccer.

Gaponga Auditory bait designed to lure fruit-eating fish.

Garota maravilhosa Marvelous wench.

Gatinha ("little [female] cat" or kitten). Teenybopper.

Gavião-belo Black-collared hawk, *Busarellus nigricollis* (Accipitridae).

Geonoma Genus of small palms of the understory, in the family Arecaceae.

Giant water lily *Victoria amazonica* (Nymphaeaceae). The plant was first named *Victoria regia* in honor of the English queen but was changed in a convulsion of botanical priority and political correctness.

Gomeleira Latex of a fig tree, *Ficus anthelminthica* (Moraceae).

Gray brocket deer *Mazama gouazoubira* (Cervidae).

Gray tinamou *Tinamus tao* (Tinamidae).

Gripe Respiratory infection, influenza.

Guaraná Understory vine, *Paullinia cupana* (Sapindaceae), native to Amazonia. The seeds are ground and made into a caffeinated beverage, the national soft drink of Brazil.

Guariba Red howler monkey, *Alouatta seniculus* (Cebidae).

Harpy eagle *Harpia harpyja* (Accipitridae), largest bird of prey in Amazonia.

Hoatzin *Opisthocomus hoazin* (Opisthocomidae).

Igarapé ("canoe trail"). Channel.

Iguana *Iguana iguana* (Iguanidae).

Imbaúba Any of several species of *Cecropia* (Cecropiaceae) trees in the western Amazon; all are indicators of disturbed areas such as riverbanks, old fields, and roadsides.

Inajá Palm, *Attalea maripa* (Arecaceae).

Inhambú acu Gray tinamou, *Tinamus tao* (Tinamidae).

Inverno Winter, which in Amazonia means the rainy season, October to April.

Ixora Genus of small trees and shrubs, often with showy red flowers, in the coffee family (Rubiaceae).

Jabutí Red-footed tortoise, *Geochelone imbricata* (Testudinidae).

Jacaré preta or *jacaré açu* Black caiman, *Melanosuchus niger* (Alligatoridae).

Jacaré tinga cascudo Dwarf caiman, *Paleosuchus palpebrosus* (Alligatoridae). During nesting season, the females often excavate a chamber inside a living termite nest, in which they lay their eggs. The termites seal the mother and her eggs behind a wall of papery excrement, leaving them isolated and unmolested until the eggs hatch.

Jacaré tinga Spectacled caiman, *Caiman crocodilus* (Alligatoridae). Before crocodilians were protected by law, this species was exported by the thousands to the United States and Europe as pets; sometimes they were sold in dime stores. It is the only palatable species of caiman; the rest have bad-smelling meat and are consumed only as starvation foods.

Jaguarundi *Herpailurus yagouaroundi* (Felidae), a lanky brown wild cat, about the size of a domestic cat.

Jambu 1) An ornamental tree, *Syzygium jambus* (Myrtaceae), beloved for its flowers and fruits; introduced from Asia, it is a close relative of the clove tree. 2) An herb, *Spilanthes acmella* (Asteraceae), prized as a vegetable, particularly in soups. It slightly anesthetizes the tongue and lips, creating a pleasant tingling sensation.

Japiim Yellow-rumped cacique, *Caricus cela* (Icteridae).

Jararaca vermelha Red fer-de-lance, *Bothrops brazili* (Viperidae).

Jauarí Spiny-trunked riparian palm, *Astrocaryum jauari*.

Jeito Savvy; street (or forest) smarts.

Kikuyu grass *Pennisetum clandestinum* (Poaceae), a grass imported from East Africa and favored by ranchers.

King vulture *Sarcoramphus papa* (Cathartidae).

Lagoa Lake.

Large-billed tern *Phaetusa simplex* (Laridae).

Leaf-nosed bat Any of the bats in the family Phyllostomidae.

Leishmaniasis Protozoal disease that creates lesions of the skin and mucous membranes, often causing the nose and palate to erode and sometimes leading to death. It is vectored by moth flies (*tatuqueira*) and reservoired in a variety of mammals, including armadillos and dogs.

Long-nosed bat *Rhynchonycteris naso* (Emballonuridae).

Louro preto Black laurel, *Mezilauris itauba* (Lauraceae), a tree whose wood is famous throughout Amazonia for its hardness, durability, and resistance to rotting, whether in the air or underwater.

Macaco barrigudo ("pot-bellied monkey"). Common woolly monkey, *Lagothrix lagothricha* (Cebidae).

Macaco zogue zogue Dusky titi monkey, *Callicebus moloch* (Cebidae).

Madrugá Dawn.

Maloca Native American communal house or longhouse.

Mamey Tree, *Pouteria sapota* (Sapotaceae), yielding a large edible fruit.

Mandi-serra Algae-scraping catfish, *Leptodoras* species (Doradidae).

Mapinguarí Evil forest spirit or animal, the size of a child, whose toes face backward; some scholars believe that it may be a ground sloth.

Maracajá Ocelot, *Felis pardalis* (Felidae).

Maracujá Passion fruit.

Mariri Native American ceremony in which hallucinogenic *caapi* is drunk.

Martim-pescador-verde Amazon kingfisher, *Chloroceryle amazona* (Alcedinidae).

Mata Forest.

Mata mata Any of several trees in the Brazil nut family (Lecythidaceae). The fibrous bark is commonly used as cigarette paper.

Mateiro 1) Woodsman. 2) Manager of a *seringal*.

Mayfly Probably *Thraulodes* species (Leptophlebiidae).

Mealy parrot *Amazona farinosa* (Psittacidae).

Midge Any of the tiny flies in the family Chironomidae.

Moon snail Air-breathing aquatic snail in the family Pulmonidae.

Morador Settler.

Moreno Brown-skinned one, a descriptive, not a pejorative, term.

Morpho butterfly Any of the twenty-five species of *Morpho* (Nymhalidae).

Motor do rabo do rato ("rat-tailed motor"). Air-cooled diesel or gasoline engine that turns a propeller at the end of a long shaft.

Mucuim Chigger, *Eutrombicula* species (Trombiculidae). About eighty species have been described in the Neotropics.

Munguba New World relative of the African baobob tree, *Pseudobombax munguba* (Bombacaceae), which bears large orange pods.

Mutuca Horsefly in the genus *Tabanus* (Tabanidae); there are about a thousand species in the Neotropics.

Mutum Amazonian razor-billed curassow, *Mitu tuberosa* (Cracidae).

Nightjar Nocturnal bird in the family Caprimulgidae.

Noctuid moth Any of the moths in the family Noctuidae.

Oaca Pellet of *buriti* palm fruit, *farinha,* and red pepper. When fed to fish, *oaca* causes them to bloat and bob to the surface.

Onça Jaguar.

Ouro branco ("white gold"). Rubber.

Paca *Agouti paca* (Agoutidae), a fawn-colored rodent.

Pacu Fish, *Myleus pacu* (Characidae), known as the silver dollar in the aquarium trade.

Patauá Palm, *Oenocarpus bataua* (Arecaceae).

Patrão Boss of a rubber estate.

Pau d'arco ("wood of the bow"). A yellow-flowering emergent tree, *Tabebuia serratifolia* (Bignoniaceae), with wood that is strong and very flexible and is therefore ideal for an archer's bow. The bark is widely used as a remedy for a variety of ailments.

Pavãozinho Sun bittern, *Eurypyga helias* (Europygidae).

Paxiúba barriguda Pot-bellied palm, *Socratea exorrhiza* (Arecaceae). The name refers to the swollen midtrunk of the mature tree.

Peconha Canvas strap that binds the feet, used to shimmy up trees.

Peixe cachorro ("dog fish"). The fish *Raphiodon vulpinus* (Characidae).

Peixe catinga ("body-odor fish"). Fish in the genus *Pimelodina* (Pimelodidae).

Peixote ("little fish"). Street urchin.

Pernilongo Black-winged stilt, *Himantopus himantopus* (Recurvirostridae).

Pescada Fish in the genus *Plagioscion* (Mugilidae).

Pica-pau-de-topete-vermelho Crimson-crested woodpecker, *Campephilus melanoleucos* (Picidae).

Pico da jaca Bushmaster snake, *Lachesis muta* (Viperidae).

Piracema Seasonal autumn migration of fish from the river into the flooded forest, often near the time of the full moon.

Piranha caju Cashew piranha, *Serrasalmus natterei* (Characidae).

Piranha preta Black piranha, *Serrasalmus rhombeus* (Characidae), the largest piranha and the only vegetarian species.

Piraqueira Oil lamp that is strapped to the forehead.

Pirarucú Arapaima gigas, the largest freshwater fish in the world, 2 meters long and weighing up to 200 kilograms, in the relictual family Osteoglossidae, the bony-tongued fishes. Its tongue has a bone covered with thousands of rasping villi, and the local people often use it as a file.

Pistoleiro Hired gunslinger; land disputes in Acre are often settled by rule of might.

Pium Biting fly in the genus *Simulium* (Simuliidae), a vector of river blindness in northern Amazonia.

Poraqué Electric eel, *Electrophorus electricus* (Electrophoridae).

Preguiça Brown-throated three-toed sloth, *Bradypus variegatus* (Bradypodidae).

Pupunha Spiny-trunked palm, *Bactris gasipaes* (Arecaceae). Like *buriti*, it is an indicator of the floodplain.

Queixada ("jawbone"). White-lipped peccary, *Tayassu pecari* (Suidae). The name refers to the animal's stout mandibles, which make a sound like snapping fingers. This is probably the way members of a pack communicate in the forest.

Raia Spotted freshwater stingray, *Potamotrygon* or *Disceus* species (Potamotrygonidae).

Railroad vine Running vines in the cosmopolitan genus *Convolvulus* (Convovulaceae), common in disturbed areas.

Rain frog *Leptodactylus* species (Leptodactylidae).

Rana False, in *lingua geral.*

Rasga mortalha ("shroud-tearer"). A nightjar, species unknown, in the family Caprimulgidae. The name refers to its rasping voice.

Real (plural, *reais*). Portuguese silver coin.

Red howler monkey *Alouatta seniculus* (Cebidae).

Red-footed tortoise *Geochelone denticulata* (Testudinae).

Regatão (plural, *regatões*). Itinerant river trader.

Restinga Ridge that runs parallel to a river.

Rio River.

Rocinha Slash-and-burn farm.

Samaúma Silk-cotton tree, *Ceiba pentandra* (Bombacaceae), a relative of the African baobabs. It is one of the tallest trees in Amazonia, growing to 60 meters.

Santo daime Hallucinogenic beverage made from the bark of *caapi* boiled with leaves of *café brava* ("wild coffee"), *Psychotria* species (Rubiaceae).

Sardinha Any of several species of silvery fish in the genus *Triportheus* (Characidae), with big pectoral fins.

Sarsaparilla Dried roots of *Smilax* species (Smilacaceae), used for flavoring.

Saúva Leaf-cutter ants in the genus *Atta* (Formicidae).

Scarab Any beetle in the family Scarabidae.

Scarlet macaw *Ara macao* (Psittacidae).

Seringal (plural, *seringais*). Rubber estate. The word is derived from *seringa,* the Native American word for rubber. The English word *syringe* (the first syringes were squirting toys made from rubber) also derives from *seringa.*

Seringueiro Rubber tapper.

Sete cores ("seven colors"). Paradise tanager, *Tangara chilensis* (Thraupinae).

Silvery marmoset *Callithrix argentata* (Callithrichidae).

Silvestre Forest.

Snail kite *Rostrhamus sociabilis* (Accipitridae), known in the United States as the Everglades kite.

Social spider Probably in the genus *Anelosimus* (Theriidae).

Sucupira preta Tree, *Sclerolobium aureum* (Caesalpiniaceae).

Sun bittern *Eurypyga helias* (Eurypygidae).

Surubim Catfish, *Pseudoplatysoma fasciatum* (Pimelodidae).

Surucuá-de-cauda-preta Black-tailed trogon, *Trogon melanurus* (Trogonidae).

Surucucú Fer-de-lance, *Bothrops atrox* (Viperidae).

Surucucú rana "False" fer-de-lance, a tree boa, *Corallus hortulanus* (Boidae).

Tambaqui Large fish (weighing up to 20 kilos), *Collosoma bidens* (Characidae), related to the piranhas.

Tapir *Tapirus terrestris* (Tapiridae).

Tarrafa Net that is cast to catch fish.

Tatajuba Hardwood tree, *Bagassa guianensis* (Moraceae).

Tatuqueira ("armadillo-lover"). Any biting moth fly in the genus *Phlebotomus* (Psychodidae), a vector of leishmaniasis. The name refers to the insect's penchant for spending the heat of the day in dark, moist armadillo burrows.

Tatuzinho ("Little Armadillo"). A cheap, popular brand of *cachaça*.

Tegu Black lizard, *Tegu tegu* (Teiidae), 1 meter long, with white or yellow markings. In English it is called the poulterer's thief because of its skill in stealing eggs from henhouses.

Terminalia Genus of trees in the family Combretaceae.

Terra firme ("firm land"). Upland that is not seasonally flooded, often low in nutrients but extravagantly rich in species.

Terra ilhada ("land-island"). Dry upland that offers refuge from floods during the rainy season.

Terra roxa do índio ("red earth of the Indian"). Anthropogenic prehistoric *terra firme* soils of high fertility resulting from the accumulation of food and waste over the centuries.

Top-knot pigeon *Ocyphaps lophotes* (Columbidae), also known as crested pigeon.

Tracajá Side-necked turtle, *Podocnemis* (Chelidae).

Traíra Voracious predatory fish, *Hoplias malabaricus* (Characidae), whose body is one-third head and mouth.

Tropeiro Screaming piha, a small but very loud bird, *Lipaugus vociferans* (Cotingidae).

Tucano-de-bico-preto Channel-billed toucan, *Ramphastos vitellinus* v. *culminatus* (Ramphastidae).

Tucuma Spiny palm with edible, scaly fruit, *Astrocaryum tucuma* (Arecaceae).

Tucunaré Fish incorrectly called ocellated bass, *Cichla ocellaris* (Cichlidae).

Tucuxí *Sotalia fluviatalis,* a member of the modern oceanic family Delphinidae. *Tucuxís,* which are only a meter long, present their triangular dorsal fins and flukes to the air before sounding; sometimes they leap clear out of the water.

Turkey vulture *Cathartes aura* (Cathartidae)

Tyrant kingbird Great kiskadee, *Pitangus sulfuratus* (Tyrannidae).

Ucuüba Tree, *Virola surinamensis* (Myristicaceae), related to nutmeg. It grows in the *várzea* from the Atlantic to the Andes. The crystallized, reddish sap of *Virola* is widely used by tribal people throughout western

Amazonia as a hallucinogenic snuff; it contains a compound similar to LSD.

Uranid moth Any of a number of species in the diurnal swallow-tailed moth family Uraniidae.

Urubu Vulture. The most common species in Amazonia are the black vulture (*Coragyps atratus*), turkey vulture (*Cathartes aura*), and king vulture (*Sarcoramphus papa*), in the family Cathartidae. All three range throughout the New World subtropics and tropics.

Vagalumi Large (up to 4 cm.) bioluminescent click beetle in the genus *Pyrophorus* (Elateridae). The celebrated explorer Alexander von Humboldt gathered them in a latticed gourd, which he used as a reading lamp.

Várzea Forest or savanna that is annually flooded with sediment-rich white water, which is often anoxic but is high in nutrients. Species diversity in the *várzea* is usually low.

Veado catinguero ("deer of secondary growth"). Gray brocket deer, *Mazama gouazoubira* (Cervidae).

Veado galeiro Virginia white-tailed deer, *Odocoileus virginianus* (Cervidae).

Verão Summer, June to September, the dry season in the Amazon.

Verme torneiro Botfly, screwworm.

Virginia white-tailed deer *Odocoileus virginianus* (Cervidae), a species with a huge range, from Canada to Peru.

Water lettuce *Pistia stratiodes* (Araceae).

Wattled jaçana *Jacana jacana* (Jacanidae), a marsh bird that does not swim but strides over aquatic vegetation on its long toes.

White-footed anopheles mosquito *Anopheles albimanis* (Culicidae), an important vector of malaria in Amazonia.

White-lipped peccary *Tayassu pecari* (Suidae).

White-winged swallow *Tachycineta albiventer* (Hirudinidae).

Wolf spider Any of a number of predatory spiders in the family Lycosidae.

Xibé Bland mixture of *farinha* and water.

Zanthoxylum Genus of trees and shrubs in the citrus family, Rutaceae.

꩜ Bibliography

Prologue

Seymour, Thomas Day. 1907. *Life in the Homeric Age.* New York: Macmillan.

2. River of Light

Cavalcante, Paulo B. 1991. *Frutas Comestíveis da Amazônia.* Belém: Graficentro/CEJUP.

Dickinson, Robert E., ed. 1987. *The Geophysiology of Amazonia: Vegetation and Climate Interactions.* New York: Wiley.

Ferraris, Carl J. 1991. *An Atlas of Representative South American Freshwater Fish Groups.* New York: Department of Herpetology and Ichthyology, American Museum of Natural History.

Janzen, D. H. 1976. "Why Bamboos Wait So Long to Flower." *Annual Review of Ecology and Systematics* 7: 347–91.

———. 1983. "Food Webs: Who Eats What, Why, How, and with What Effects in a Tropical Forest?" In F. B. Golley, ed., *Tropical Rain Forest Ecosystems.* Amsterdam: Elsevier.

Salati, Eneas, Wolfgang J. Junk, Herbert O. R. Shubart, and Adélia Engrácia de Oliveira. 1983. *Amazônia: Desenvolvimento, Integração e Ecologia.* São Paulo: Conselho Nacional de Desenvolvimento Científico e Tecnologico.

Salati, Eneas, and Peter B. Vose. 1984. "Amazon Basin: A System in Equilibrium." *Science* 225: 129–38.

Salati, Eneas, Peter B. Vose, and Thomas E. Lovejoy. 1986. "Amazon Rainfall, Potential Effects of Deforestation, and Plans for Future Research." In Ghillean T. Prance, ed., *Tropical Rain Forests and the World Atmosphere,* pp. 61–74. American Association for the Advancement of Science, Selected Symposium 101. Boulder, Colo.: Westview Press.

Silva, Marlene Freitas da, Pedro Luiz Braga Lisbôa, and Regina Célia Lobato Lisbôa. 1977. *Nomes Vulgares de Plantas Amazônicas.* Manaus: Instituto Nacional de Pesquisas da Amazônia.

Smith, Nigel J. H. 1979. *A Pesca no Rio Amazonas.* Manaus: Instituto Nacional de Pesquisas da Amazônia.

van den Berg, Maria Elisabeth. 1982. *Plantas Medicinais na Amazônia; Contribuição ao seu Conhecimento Sistematico.* Belém: Museu Paraense Emilio Goeldi.

3. An Expedition of Poets

Bastos, A. C. Tavares. 1866. *O Vale do Amazonas: a Livre Navegação do Amazonas, Estatística, Produções, Comércio, Questões Fisicais do Vale do Amazonas.* São Paulo: Companhia Editora Nacional (1975).

Coates, Austin. 1987. *Commerce in Rubber: 250 Years.* New York: Oxford University Press.

Coêlho, Enice Mariano. 1982. *Acre: o Ciclo da Borracha (1903 – 1945).* Niterói: Instituto de Ciências Humanas e Filosofia, Centro de Estudos Gerais, Universidade Federal Fluminense.

Condamine, Charles-Marie de la. 1747. *Voyage Sur L'Amazone.* (Choix de Textes.) Paris: François Maspero (1981).

Cunha, Euclides da. 1986. *Um Paraíso Perdido; Ensaios, Estudos e Pronunciamentos Sobre a Amazônia.* Rio de Janeiro: José Olympio Editora.

Hay, James de Vismes Drummond. 1870. "Report on the Industrial Classes in the Provinces of Pará and Amazonas, Brazil." Appendix to Wickham, Henry Alexander. 1872. *Rough Notes of a Journey through the Wilderness from Trinidad to Pará, Brazil, by Way of the Great Cataracts of the Orinoco, Atabapo, and Rio Negro to James de Vismes Drummond Hay, C.B., H.B.M., Consul for Valparaiso (Late of Pará),* pp. 291–301. London: W.H.J. Carter.

Hecht, Susanna, and Alexander Cockburn. 1989. *The Fate of the Forest: Developers, Destroyers and Defenders of the Amazon.* New York: Verso.

Hepper, F. Nigel. 1982. *Kew: Gardens for Science and Pleasure.* London: Her Majesty's Stationery Office.

Lima, Manoel Ferreira. Undated. *O Acre: Seus Aspectos Físicos e Geográficos, Só-cio-Econômicos, Historicos e Seus Problemas,* terceiro edição. Rio Branco.

Lopes, Orlando Corrêa. 1906. *O Acre e O Amazonas.* Rio de Janeiro: Jornal do Commercia.

Melo, Hélio. 1985. *As Experiêncas do Caçador: Do Serengueiro para o Serginueiro.* Rio Branco: Fundação Cultural do Acre.

Rancy, Cleusa Maria Damo. 1986. *Raízes do Acre (1870–1912).* Rio Branco: Secretaria de Estado de Educação e Cultura, Governo do Estado do Acre.

Reis, A.C.F. 1953. *O Seringal e O Seringueiro.* Rio de Janeiro: Servico de Informação Agrícola, Ministerio da Agricultura.

Spruce, Richard. 1908. *Notes of a Botanist on the Amazon and Andes.* London: Macmillan.

Tambs, Lewis A. 1966. "Rubber, Rebels, and Rio Branco: The Contest for Acre." *Hispanic American Historical Review* 46(3): 254–73.

Thephilo, Rodolfo. 1883. *História da Seca do Ceará.* Fortaleza: Typ. do Libertador.

Tocantins, Leandro. 1961. *Formação Histórica do Acre,* vols. 1–3. Rio de Janeiro: Conquista.

Weinstein, Barbara. 1983. *The Amazon Rubber Boom 1850–1920.* Stanford, Calif.: Stanford University Press.

Wickham, Henry Alexander. 1872. *Rough Notes of a Journey through the Wilderness from Trinidad to Pará, Brazil, by Way of the Great Cataracts of the Orinoco, Atabapo, and Rio Negro to James de Vismes Drummond Hay, C.B., H.B.M., Consul for Valparaiso (Late of Pará).* London: W.H.J. Carter.

Wisniewski, Alfonso. 1983. *A Boracha na Sócio-Economia do Estado do Pará.* Belém: Ministerio da Educação e Cultura.

Wolf, Howard, and Ralph Wolf. 1936. *Rubber; a Story of Glory and Greed.* New York: Colvici, Friede.

Woodroffe, Joseph F. 1915. *The Rubber Industry of the Amazon, and How Its Supremacy Can Be Maintained.* Oxford: John Bale, Sons, and Danielsson.

4. River of Hunger

Medem, Federico. 1983. *Los Crocodylia de Sur America,* vol. 2. Bogotá: Universidad Nacional de Colombia.

Meggers, Betty J. 1971. *Amazonia: Man and Culture in a Counterfeit Paradise.* Chicago: Aldine.

5. *What the Bees Said*

Balée, W., and D. G. Campbell. 1989. "Ecological Aspects of Liana Forest, Xingu River, Amazonian Brazil." *Biotropica* 22(1): 36–47.

Smith, N. 1980. "Anthrosols and Human Carrying Capacity in Amazonia." *Annals of the Association of American Geographers* 70(4): 554–66.

6. *On the Line*

Ayres, J. M., and T. H. Clutton-Brock. 1992. "River Boundaries and Species Range Size in Amazonian Primates." *American Naturalist* 140(3): 531–37.

Balée, William. 1994. *Footprints of the Forest.* New York: Columbia University Press.

Bates, Henry Walter. 1854. *The Naturalist on the River Amazons.* London: John Murray.

Campbell, D. G., J. L. Stone, and A. Rosas, Jr. 1992. "A Comparison of the Phytosociology of Three Floodplain (*Várzea*) Forests of Known Ages, Rio Juruá, Western Brazilian Amazon." *Botanical Journal of the Linnaean Society* 108: 213–37.

Connell, J. H. 1978. "Diversity in Tropical Rain Forests and Coral Reefs." *Science* 199: 1302–9.

Delslow, J. S. 1987. "Tropical Rain Forest Gaps and Tree Species Diversity." *Annual Review of Ecology and Systematics* 18: 431–51.

Denevan, W. M. 1992. "The Pristine Myth: The Landscape of the Americas in 1492." *Annals of the Association of American Geographers* 82(3): 369–85.

———. 1996. "A Bluff Model of Riverine Settlement in Prehistoric Amazonia." *Annals of the Association of American Geographers* 86(4): 654–81.

Huston, M. A. 1994. *Biological Diversity.* New York: Cambridge University Press.

Prance, Ghillean T., ed. 1982. *Biological Diversification in the Tropics.* New York: Columbia University Press.

Salo, Jukka, Risto Kalliola, Ilmari Häkkinen, Yrjö Mäkinen, Pekka Niemelä, Maarit Puhakka, and Phyllis D. Coley. 1986. "River Dynamics and the Diversity of Amazon Lowland Forest." *Nature* 322: 254–58.

Tuomisto, Hanna, Kalle Ruokolainen, Risto Kalliola, Ari Linna, Walter Danjoy, and Zoila Rodriguez. 1995. "Dissecting Amazonian Biodiversity." *Science* 269: 63–66.

Wallace, Alfred Russel. 1891. *Tropical Nature*. London: Macmillan.

———. 1895. *A Narrative of Travels on the Amazon and Rio Negro with an Account of the Native Tribes and Observations on the Climate, Geology, and Natural History of the Amazon Valley*. London: Ward, Lock and Bowden.

Webb, S. David. 1995. "Biological Implications of the Middle Miocene Amazon Seaway." *Science* 269: 361–62.

Whitmore, T. C., and G. T. Prance, eds. 1987. *Biogeography and Quaternary History in Tropical America*. Oxford: Clarendon Press.

Young, Truman P., and Victoria Perocha. 1994. "Treefalls, Crown Asymmetry, and Buttresses." *Journal of Ecology* 82: 319–24.

7. The Naming of Parts

Bentham, G. 1875. "Revision of the Suborder Mimoseae." *Transactions of the Linnaean Society of London*. 30(3): 600–632.

Malhi, Y., and O. L. Phillips. 2004. "Tropical Forests and Global and Atmospheric Change: A Synthesis." *Philosophical Transactions: Biological Sciences* 359: 549–55.

Mayr, Ernst. 1992. "Species Concepts and Their Application." In Marc Ereshefsky, ed., *The Units of Evolution, Essays on the Nature of Species*, pp. 15–26. Cambridge: Massachusetts Institute of Technology Press.

Morais, Paula B., Marlúcia B. Matrins, Louis B. Klaczko, Leda C. Mendonça-Hagler, and Allen N. Hagler. 1995. "Yeast Succession in the Amazon Fruit *Parahancornia amapa* as Resource Partitioning among *Drosophila* spp." *Applied and Environmental Microbiology* 61(12): 4251–57.

Pennington, T. D. 1997. *The Genus* Inga: *Botany*. London: Royal Botanic Garden, Kew.

Richardson, James E., R. Toby Pennington, Terence D. Pennington, and Peter M. Hollingsworth. 2001. "Rapid Diversification of a Species-Rich Genus of Neotropical Rain Forest Trees." *Science* 293: 2242–45.

8. The Gift

Dixon, James R., and Pekka Soini. 1986. *The Reptiles of the Upper Amazon Basin, Iquitos Region, Peru.* Milwaukee: Milwaukee Public Museum.
Silva, Marcelo, Jr. 1956. *O Ofidismo no Brasil.* Rio de Janeiro: Serviço Nacional de Educação Sanitaria.

9. River of Terns

Bates, Henry Walter. 1854. *The Naturalist on the River Amazons.* London: John Murray.
Diamond, Jared. 1986. *Guns, Germs, and Steel.* New York: Norton.
Edwards, William H. 1847. *A Voyage up the River Amazon, Including a Residence at Pará.* New York: D. Appleton.
Fritz, Samuel. 1922. *Journal of the Travels and Labours of Father Samuel Fritz in the River of the Amazons between 1686 and 1723.*
Gibbon, Lardner. 1854. *Exploration of the Valley of the Amazon Made Under Direction of the Navy Department.* Washington, D.C.: A.O.P. Nicholson.
Hemming, John. 1978. *Red Gold: The Conquest of the Brazilian Indians 1500 – 1760.* Cambridge: Harvard University Press.
———. *Amazon Frontier.* Cambridge: Harvard University Press.
Herndon, William. 1854. *Exploration of the Valley of the Amazon.* Ed. Hamilton Basso (1952). New York: McGraw-Hill.
Parreira, José Otávio. 1991. *Atlas Geográfico Ambiental do Acre.* Rio Branco: Secretario de Meio Ambiente do Acre.
Porro, Antonio. 1981. "Os Omáguas do alto Amazonas: demografia e padrões de povoamento no século XVII." In *Contribuiçoes à Antropologia em Homenagem ao Professor Egon Schaden,* pp. 207–321. São Paulo: Coleção Museu Paulista, Série Ensaios 4.
———. 1994. "Social Organization and Political Power in the Amazon Floodplain: The Ethnohistorical Sources." In Anna Roosevelt, ed. *Amazonian Indians from Prehistory to the Present,* pp. 79–94.
Ross, Eric. R. 1978. "The Evolution of the Amazon Peasantry." *Journal of Latin American Studies* 10(2): 193–218.
Spruce, Richard. 1908. *Notes of a Botanist on the Amazon and Andes.* London: Macmillan.

Whitehead, Neil Lancelot. 1994. "The Ancient Amerindian Polities of the Amazon, the Orinoco, and the Atlantic Coast: A Preliminary Analysis of Their Passage from Antiquity to Extinction." In Anna Roosevelt, ed., *Amazonian Indians from Prehistory to the Present*, pp. 33–53.

10. A Land of Ghosts

Chandless, W. 1867. "A Voyage of Exploration on the Rio Juruá." Unpublished journal.

———. 1869. "Notes of a Journey up the River Juruá." *Journal of the Royal Geographical Society* 39: 296–311.

Kazin, Alfred. 1988. *A Writer's America: Landscape in Literature*. New York: Knopf.

Lathrup, Donald W. 1970. *The Upper Amazon*. London: Thames and Hudson.

Mendonça, Belarmino. 1905. *Memoria da Commissão Mixta Brasileiro—Peruana de Reconhecimento do Rio Rio Juruá*. Rio de Janeiro: Imprensa Nacional.

Métraux, Alfred. 1948. "Tribes of the Jurua-Purus Basins." In Julian H. Steward, ed., *Handbook of South American Indians*, vol. 3: *The Tropical Forest Tribes*, pp. 657–712. Washington, D.C.: United States Government Printing Office.

Neto, Carlos de Araújo Moreira. 1988. *Indios da Amazônia, de Mairoia a Minoria (1750—1850)*. Petrópolis: Editora Vozes.

Nugent, Stephen. 1994. "Invisible Amazonia and the Aftermath of Conquest: A Coda to the Quincentenary Celebrations." *Journal of Historical Sociology* 7(2): 224–41.

Reis, A. 1931. *História do Amazonas*. Manaus: Officinas Typográphicas de A. Reis.

Ribeiro, D. 1967. "Indigenous Cultures and the Languages of Brazil." In J. Hopper, ed., *Indians of Brazil in the 20th Century*. Washington, D.C.: Institute of Cross-Cultural Research.

Rivet, Paul. 1920. "Les Katukina, étude linguistique." *Journal de la Société des Américanistes de Paris* 12: 283–89.

Rivet, Paul, and Constant Tastevin. 1919–1920. "Les langues du Purús, du Juruá et des régions limitrophes. 1. le groupe arawak pré-andin." *Anthropos* 14–15: 857–90.

————. 1921. "Les tribus indiennes du Purús, du Juruá et des régions limitrophes. I. le groupe arawak pré-andin." *La Geografie* 35: 449–482.

————. 1921–1922. "Les langues du Purús, du Juruá et des régions limitrophes. I. le groupe arawak pré-andin." *Anthropos* 16–17: 298–325,
819–28.

————. 1923–1924. "Les langues du Purús, du Juruá et des régions limitrophes. I. le groupe arawak pré-andin." *Anthropos* 18–19: 104–13.

————. 1938. "Les langues arawak du Purús et du Juruá (groupe arauá)."
Journal de la Société des Américanistes de Paris 30: 71–114, 235–88.

Roosevelt, Anna. 1994. "Amazonian Anthropology: Strategy for a New Synthesis." In Anna Roosevelt, ed., *Amazonian Indians from Prehistory to the
Present: Anthropological Perspectives,* pp. 1–29. Tucson: University of Arizona Press.

Steward, Julian H., and Alfred Métraux. 1948. "Tribes of the Peruvian and
Ecuadorian Montana." In Julian H. Steward, ed., *Handbook of South
American Indians,* vol. 3: *The Tropical Forest Tribes,* pp. 525–656. Wash-
ington, D.C.: United States Government Printing Office.

Tastevin, Constant. 1920. "Quelques considérations sur les Indiens de Juruá."
Bulletins et Mémoires de la Société du Anthropologie de Paris. 10: 144–54.

————. 1924a. "Les études ethnografiques et linguistiques du P. Tastevin en
Amazonie." *Journal de la Société des Américanistes de Paris* 16: 421–25.

————. 1924b. "Chez les Indiens du haut Jurua." In *Les Missions Catholiques,*
pp. 56, 65–67, 78–80, 90–93, 101–4. Lyon.

————. 1924c. "Le fleuve Juruá." *La Géographie* 33(1): 1–22.

————. 1928. "Sur les Indiens Katukina." *L'Ethnographie* 17–18: 130–32.

II. Angelim

Kullberg, Mary. 1993. *Morning Mist, Thoreau and Basho Through the Seasons.*
Tokyo: Weatherhill.

Moura, Pedro de, and Alberto Wanderly. 1938. *Noroeste do Acre, Reconheci-
mentos Geologicos para Petroleo.* Rio de Janeiro: Serviço do Fomento da
Produção Mineral, Departamento Nacional da Produção Mineral, Ministerio da Agricultura, Republica dos Estados Unidos do Brasil.

Schopf, W., ed. 1983. *Earth's Earliest Biosphere.* Princeton: Princeton University Press.

ꙮ Index

Index

Araras (scarlet macaws), 71,
72. *See also* Scarlet
macaws
Araua Indians, 178
Archaeological Institute of
Pernambuco, 222n.12
Archaeology, 171, 173
Archean epoch, 207–8,
228n.4
Arito (expedition member).
See Rosas, Arito, Jr.
Arlindo da Fonte, Manoel
(husband of Dona
Cabocla), 98–100,
150
Armadillo, 110
Army ants, 5. *See also* Ants
Arpão (harpoon), 85
Arpoador (harpoonist), 85
Aruanã fish, 45, 157
Aspidosperma tree, 127
Assacú logs, 20, 78–79
Astor, John Jacob, 63
Atlantic Ocean, water va-
por from, 40–41,
219–20n.2
Ausira, Dona, 190–94, 198,
206, 214
*photograph, following
p. 105*
Australia, 206–8
Aviação (grubstake), 52, 74,
77, 101
Ávila e Silva, Francisco
de, 67
Azevedo, Taumaturgo
de, 67
Aztecs, 221n.2

Babassu palm, 224n.4
Bacteria, 9, 107, 146
Bacurãos (nightjars), 88
Bacuri trees, 132
Bairro flutuante (shanty-
town), 20–21, 24, 78–
79, 213
Ball court, 221n.2
Bamboos, 32, 34, 41–42,
220n.3

Banana trees (*bananeiras*)
and bananas, 32, 61,
78, 92, 94, 96, 98,
128–29, 155, 183, 189,
222n.3, 224n.3
Banisteriopsis vine, 60
Banks, Joseph, 222n.17
Barbados, 162–63
Barbosa de Lima, Fran-
cisco Xavier. *See*
Chiquininim
Barges, 15–16, 24
Barracão (rubber tappers'
dormitory), 49, 51,
59–60, 64, 68, 76–
77, 149, 180, 189, 212
Barracão Aurora, 28, 172–73
Barter system, 52, 175
Batedores (professional Indian hunters), 60–61
Bates, Henry Walter, 120–
21, 168–69, 223n.17
Batista, Edimar Ferreira
(Edinho, expedition
member), 26, 34
Bats, 4, 17, 32, 36, 45, 81,
82, 102, 104, 130, 139,
144, 156, 157, 223n.1
Beaches, 23, 36, 43, 181,
211
Beans, 42, 79, 94, 98, 167
Bees, 4, 36, 39, 90, 91, 99,
122, 123, 132, 137, 139,
209
Beetles, 4, 21, 32, 45, 90,
110, 129, 130, 156, 157
Beija-flor-de-banda-branca
(hummingbird), 128
Belém (formerly Pará), 55,
56, 57, 59, 61, 63, 64,
66, 161, 162, 166, 167,
168, 175, 200, 227n.5
Belgian Congo Company,
63
Belo Horizonte, 72
Benjamin Constant,
220n.2
Bentham, George, 142
Beri-beri, 182

Bico de argulha (bluish-
fronted jacamar), 133,
225n.3
Biological diversity, 7, 12,
16, 108, 116, 119–22,
222–23n.17
Biomass, 107–8
Birds, 34–35, 51–52, 81, 89–
90, 129, 151. *See also
entries for specific birds*
Bitterns, 71
Bladderworts, 38, 39
Bligh, William, 222n.17
Boa, 45, 82–83
Bolas (balls of smoked rub-
ber), 14, 23, 53, 75,
76–77, 100, 101
Bolivia, 62–63, 66, 142
Bolivian Syndicate, 63–64
Bolota (Francisco José
de Souza), 195–97,
213–14
Bom Sossego, 28
Bordellos, 19, 29, 149–50
Botany and botanists, 3–
7, 11–12, 26–28, 184.
See also Expedition;
Scientific studies
of tropical forest;
Taxonomists
Botflies, 51–52, 90
Botinhos (catfish), 43, 44
Bôtos vermelhos (red dol-
phin), 80, 89
Bouguer, Pierre, 54–55
Brancos, 189, 191
Brazil, 62–69. *See also*
Acre; Amazon River
Valley; Cruzeiro do
Sul; Expedition; Na-
tive Americans; Rub-
ber tapping; Trans-
amazon Highway;
Tropical forest
Brazilian Commission for
Reconnaissance of the
Juruá, 171–72, 180–81
Brocket deer, 146, 224n.5.
See also Deer

CPSIA information can be obtained
at www.ICGtesting.com
Printed in the USA
LVOW03s0256041017
551042LV00011B/278/P

9 780813 540528